tredition

tredition was established in 2006 by Sandra Latusseck and Soenke Schulz. Based in Hamburg, Germany, tredition offers publishing solutions to authors and publishing houses, combined with worldwide distribution of printed and digital book content. tredition is uniquely positioned to enable authors and publishing houses to create books on their own terms and without conventional manufacturing risks.

For more information please visit: www.tredition.com

TREDITION CLASSICS

This book is part of the TREDITION CLASSICS series. The creators of this series are united by passion for literature and driven by the intention of making all public domain books available in printed format again - worldwide. Most TREDITION CLASSICS titles have been out of print and off the bookstore shelves for decades. At tredition we believe that a great book never goes out of style and that its value is eternal. Several mostly non-profit literature projects provide content to tredition. To support their good work, tredition donates a portion of the proceeds from each sold copy. As a reader of a TREDITION CLASSICS book, you support our mission to save many of the amazing works of world literature from oblivion. See all available books at www.tredition.com.

 Project Gutenberg

The content for this book has been graciously provided by Project Gutenberg. Project Gutenberg is a non-profit organization founded by Michael Hart in 1971 at the University of Illinois. The mission of Project Gutenberg is simple: To encourage the creation and distribution of eBooks. Project Gutenberg is the first and largest collection of public domain eBooks.

'Murphy' A Message to Dog Lovers

Ernest Gambier-Parry

Imprint

This book is part of TREDITION CLASSICS

Author: Ernest Gambier-Parry
Cover design: Buchgut, Berlin – Germany

Publisher: tredition GmbH, Hamburg - Germany
ISBN: 978-3-8472-1528-8

www.tredition.com
www.tredition.de

Copyright:
The content of this book is sourced from the public domain.

The intention of the TREDITION CLASSICS series is to make world literature in the public domain available in printed format. Literary enthusiasts and organizations, such as Project Gutenberg, worldwide have scanned and digitally edited the original texts. tredition has subsequently formatted and redesigned the content into a modern reading layout. Therefore, we cannot guarantee the exact reproduction of the original format of a particular historic edition. Please also note that no modifications have been made to the spelling, therefore it may differ from the orthography used today.

"And when his son had prepared all things for the journey, Tobit said, Go thou with this man, and God, which dwelleth in Heaven, prosper your journey, and the angel of God keep you company. So they went forth both, and the young man's dog with them."

TOBIT V. 16.

"MURPHY"

His dog.
by Him.

"MURPHY"

A MESSAGE TO DOG-LOVERS

BY

MAJOR GAMBIER-PARRY

With two drawings by the author

NEW YORK
MITCHELL KENNERLEY
1913

COPYRIGHT 1913 BY
MITCHELL KENNERLEY

PRINTED IN AMERICA
TO
THAT VAST HOST IN THE HUMAN FAMILY
THAT LOVES DOGS
AND THAT INCLUDES WITHIN ITS RANKS
THE GOOD, THE GREAT, AND THE INSIGNIFICANT

THESE PAGES

ARE RESPECTFULLY INSCRIBED
BY ONE OF
THE COMMON RANK AND FILE

ILLUSTRATIONS

"HIS DOG"

"ALAS!"

"MURPHY"
A MESSAGE TO DOG-LOVERS

I

Yes. He was born in the first week of June, in the year 1906. Quite a short while ago, as you see—that is, as we men count time—but long enough, just as a child's life is occasionally long enough, to affect the lives—ay, more, the characters—of some who claimed to be his betters on this present earth, with certainties in some dim and distant heaven that might or might not have a corner here or there for dogs.

His parentage was that of a royal house in purity of strain and length of 2 pedigree, and he first saw the light in the yard of a mill upon the river, where the old wheel had groaned for generations or dripped in silence, according as the water rose or fell, and corn came in to be ground.

There were others like him in appearance in the yard; on the eyot on which the mill-buildings stood, gorgeous in many-coloured tiles; round the dwelling-house, or in a large wired enclosure close by. His master, the Over-Lord, bred dogs of his kind for the nonce, not necessarily for profit, but because, with a great heart for dogs, he chose to, claiming indeed the proud boast that not a single dog of his class walked these Islands that was not of his strain—and claiming that, moreover, truly.

At one period there might have been counted, in and around this mill-yard, no less than thirty-eight dogs, young and 3 middle-aged, and all more or less closely related. But while this number was much above the average, the congestion that arose thereby was chargeable with the single unhappy episode in Murphy's life, concerning which he often spoke to me in after days, and the effect of which he carried to the end. Of this, however, more later.

Life in the midst of such a company—Irishmen all—necessarily meant a more or less rough-and-tumble existence, where the strongest had the best of it, and the weaker ones were knocked out, when the Master was not there to interfere. Each one had to find his own level by such means as he could, and thus this great company, or school, of dogs resembled in many particulars those other schools to which We are sent Ourselves, or send those other sons of Ours. The training to be got here, as elsewhere, developed primarily, indeed, and all unconsciously, the first and greatest of requisites in life, whether for dog or man. And if, in some instances, evil characteristics, such as combativeness, selfishness, and the habit of bad language, became accentuated, in spite of the stern discipline of the place, their opposites—good temper, a light and happy disposition, and a civil tongue—received their meed of recognition even from the bigger fellows, like Pagan I. or II., or that Captain of the School, often spoken of with bated breath—Postman, Murphy's father, mated afterwards to the great beauty, Barbara, both being of the bluest of blue blood.

The young were taught their place, and that further quality, now dropping out of fashion—how to keep it. Or each one had a lesson in yet another virtue, still more out of date, being judged no longer necessary or becoming in this very modern world, and as only showing a silly deference if exhibited at all. Respect was, in truth, the chief of all virtues here inculcated—respect for age, for old dogs are no longer to be challenged; respect for strength and the great unwritten laws; respect for sex; respect for those who had shown themselves to be the better men; respect for such as neither fought nor swore but held their own by character alone.

It was, for instance, not correct for the young to approach the older members of the school and claim equality, for, strange as it may seem, equality had no place here, save that all were dogs. Nor when a bigger fellow had a bone, won, earned, or come by of his own enterprise, was it deemed fitting that the young should do more than watch at respectful distance, with ears drooped and envy curbed as well as might be. By such methods the meaning of the sacredness of property was taught; and also, that without due regard to this last there could be security for no one, or for anything that he might own.

True, some of this company here, suffering from swelled-head, the harebrained impetuosity of youth, or judging that to them alone had been bequeathed the secret of all requisite reforms, advanced theories of their own composing. Of course they found adherents, especially when gain was scented, for to profit at another's expense is not unpopular, in some directions, from the top to the bottom of the world. But, as a rule, these theories were not long-lived. The company, so to speak, found themselves, and the innate good sense they claimed to have came to their aid, before the whole school was set generally by the ears, or 7 the Over-Lord was called upon to interfere.

Thus, where a fellow's own was concerned the cry with the really honest was, "Hands off, there!"—blood being rightly spilt, if necessary, in defence thereof, as it always will be, till the last of dogs and men lie down and die. Of course if one or other left his own unguarded, or, overcome by plethora, fell asleep, or grew fat and careless, then another of his standing came and took that property away. In such an event, he who had lost could do no more than whimper cur-like, while those lying round the yard would look up to see what the shindy was about, and then quietly remark, "*That's* as it should be."

Then again, when, on a sultry afternoon in this first summer of Murphy's life, some older members of the family betook themselves to such cool places on 8 the eyot as the shadows cast by the wide eaves of the mill, it was ordered they were to be left in peace and not plagued by younger folk, however good-natured they might be. Nor were others to be followed when they stole away to the opening of the mill-race—where the water came out at speed, brown and foaming, from the dark shadows under the floors—to listen, maybe, half asleep, to the great wheel groaning its solemn music, as the dripping green paddles threw off a cool mist to refresh the jaded air.

However strange such a choice might seem to those of restless spirit, it was not more so than that of others who, careless of themselves, preferred a hole in the dust of the upper yard among the Buff Orpingtons, and the grilling heat of the midsummer sun. There must be differences of taste here as elsewhere. The 9 spot chosen must be respected, not only because it was the home for the time, however short, but also because here was privacy, and it was not right that such should be at any time invaded, if rightly and obviously sought—at least, so was it judged by those who inhabited the island at this period.

That Murphy noticed all these things goes without saying. He kept them mostly to himself, after the manner of his kind; but he watched nevertheless closely, his black eyebrows moving continually just above his eyes, as he lay in the rough grass in the shade of the pollard willows, or beneath the whispering aspens.

At this time he had not long emerged from the limp stage, when hind-quarters would continually give way, and there was nothing to be done but rest on one haunch and try to look wise, being continually 10 bothered by the flies. After a while he began to grow stronger and more comely, his ears darkened, and his eyes—put in, as they say, with a dirty thumb—grew larger, taking on that exceeding brightness that made passers-by look and look again. He was also allowed further afield when his turn came. There were walks along the river-banks, in company with half-a-dozen of the others; and before he was six months old he could run a good distance with a horse and trap, ere he would come to the step and look up with a laugh, saying, "Here, take me up; I'm blown!" The old horse in the shafts knew the ways of the dogs well, and would shorten his pace, and indeed pull up altogether, if a thoughtless one was likely to be injured. It was probably from this that Murphy suffered all his life from a mistaken notion that it was the duty of 11 horses, as well as drivers of all kinds, to get out of his way, and not he necessarily out of theirs.

It was a happy life in a land of happiness and freedom, though discipline was stern, and all had to pass their period of training. Sooner or later each one was judged upon his merits, as well by his comrades as by the great, tall Over-Lord, to whom primarily they owed allegiance. And if such judgment was occasionally fallacious, as it frequently is, the world over, when based upon such points alone, it worked out fairly when the time arrived for an estimate to be made of the character that every one here was entitled to—when the first home had to be left behind, and the world faced in town or country, up or down the greater river of a common life.

For such a temperament as Murphy's, a life like this was happiness itself. He 12 was sociable, and loved company intensely, though preferably the company of Man. Solitude he abhorred; games were his delight; for killing things, even were it a rat from one of the thousand holes he met with when walking by the river, he never cared, and indeed appeared never quite to understand. "Live and let live" was his motto, while playing always the game of "catch-who-catch-can."

There was no reason to bring pain into the field at all. Life to him was a condition full of smiles, or to be made so, though there was snarling round the corner, as well as folk of difficult temperament to remain puzzlers to the end. Those about, therefore, were to be reckoned friends, and to be met in such way as better dogs themselves lay down. Their society obviously had its rules, which, if occasionally broken, were yet 13 to be known and recognised, just as they themselves, though dogs, were able to discern that the members of that other society, on to which they were apparently grafted, had theirs.

These last and they themselves were nothing less than partners—so it seemed to him—in a great game, to be played always in good heart and with the spirit of true sportsmanship. Both moved according to law, the only difference between the two being that Men held the power of the Veto—and exercised it too often, he would add in his perfect, well-bred manner, in a way that declared their ignorance. Men, he averred, would always insist on assuming that their laws were right at all times, and, furthermore, were always applicable to dogs, forgetting that, more often even than themselves, dogs were moved by laws imperious. 14

Had he been as the majority of dogs, he would, when such thoughts occupied his brain, have joined no doubt unhesitatingly in Puck's song—

"Lord, what fools these mortals be!"

But, then, this is where he differed from that majority. Man was his friend. Friendship meant loyalty, and loyalty should be unstained.

There was much in what he said. On many an occasion a dog will show that he knows better than a man, and can do things that transcend Man's boasted powers. We all know that—or should do so—

for the moment may arrive when we find ourselves dependent on the judgment of a dog. To fail to recognise it then is to create difficulties and to blunder badly, causing the most tractable of our friends to look up with a puzzled expression in their eyes, and the more head-strong and outspoken to go ahead, with this sentence, flung back over the shoulder—"*You fools, you; when will you understand!*"

And the fun of it all is that Man with his self-assurance, and that limited vision of his of which he seems sometimes completely unaware, thinks that he is training the dog, whereas the dog is perfectly capable, as will be shown, of at least in some directions training him. Thus, where differences arise, Man jumps to his conclusions and claims his prerogative. It is a sorry business when an all-too-hasty punishment follows, as it often does, for Man—so Murphy used to say—would find himself very frequently to be wrong. But then Murphy, when he talked like this in the after days, showing how easily We might make mistakes, and explaining so much that was not wholly realised before, caused sundry folk to wonder whether in some previous life he had in his spare time studied Bentham. For dogs or men to make mistakes is not necessarily for them to do wrong. "To trace errors to their source is often to refute them."

He often quoted that; but on the only occasion on which he was asked about his previous studies he remained silent. He and his Master were sitting on the hillside, far away from the hum of men—as, in fact, they mostly were. His eyes were ranging over the valley to the skyline. "That's the way to look, my dear master," he appeared to be saying—"that's the way to look. Never run heel way. For you and me there is a future. Look ahead, and cast forward; never look behind!"

His remarks often, in this way, touched lightly on great questions.

II

To look ahead in the hey-day of youth is to look forward to unclouded happiness. And, no doubt, to Murphy and those of his own age, the fact that the summer waned and that autumn followed, when leaves fell mysteriously from the trees and there were sporting scents in the air, made little difference to their outlook. Happiness had no relation to the seasons: they were all good in their turn. Jolly times ranged from spring to winter. And, perhaps, winter after all was best.

It was on a winter day, in fact, that Murphy first made a mark in the mind of his Over-Lord, and it came about like this.

The day before had been typical of late January. The sun had not shone 18 since daybreak. The sky to the north was lead colour, and the wind was blowing through snow. If it froze on the north side of the hedgerows, it thawed on the south—the coldest condition of all.

There were covered places for the dogs of the mill, with plenty of straw, and when one or two who had been out for a walk came in and said there would be snow before another morning dawned, those who heard the remark curled themselves tighter or drew closer to their more intimate friends. And as they slept and woke, and slept again, they saw the lights go out one by one, save those in the mill itself, for barges had come with loads of grain, and the mill was working all night. They could hear the steady "throb," "throb" of the great mill-wheel and the plash of the distant waters; but just before the new dawn 19 these sounds gave way to a hum that played a muffled music in the trees. The men's footsteps never sounded at all, till they were close at hand; and then the mill slowly stopped as though tired, and silence reigned supreme in the cold. Dogs and men slept firmly for a little: Nature was at work putting a new face upon the world.

And after all that there followed the joyousness of a cloudless morning, as the stars faded out, and the pale sun lit up a world that was now pure white. Snow lay everywhere to the depth of three inches—not more—for it had spread itself evenly in the stillness, and covered the ground, and the roofs, and the barges that had come with the grain, making everything look strange, even to the waters that were licking the banks, and that somehow or other had turned the colour of green bottle-glass.

Then, by-and-by, came the Over-Lord, and called this name and that; and the last that he called was "Murphy."

Here were games indeed! Here was something new to play with; to be skipped and rolled and gambolled in to heart's content; to be even bitten at, and swallowed till forbidden. Why, this new material that the younger ones had never seen before called even the limpest to forget his limpness, as though new blood flowed in his veins and he were endowed with a new life!

They were soon out of the yard, and away down the lane. And then the Over-Lord turned into the fields and struck a right-of-way that led in direction of a hamlet two miles distant. Here many of the meadows were thirty acres and over in extent, flat as any floor, with great elm trees in their hedgerows. They were untenanted now by sheep or cattle, for these had been driven off the night before to higher ground, by men who kept an eye upon the weather. The virgin surface of the snow lay glittering gold and silver in the early morning sun, with here and there, as a contrast, the long shadows of the limbs of a great oak or elm, cast as though some one had traced its pattern for fun with a brushful of the purest cobalt.

There were only five dogs out that morning. Three were now fastened to a leash; one other was very old, and he and Murphy were allowed what latitude they liked. So presently it chanced that Murphy found himself some way from the rest, and suddenly called upon to show what he could do. As he went, he came upon a slight rise in the snow, as though something lay beneath. The more experienced would have known what that was, for their noses would have told them in a trice. When snow falls and a hare finds itself being gradually covered by the flakes, it does what it can to bury itself deeper; but always with this eye on life — that it assiduously keeps a hole open that it may breathe, and always to the leeward. Such is one of many evidences of clever instinct to be met with for ever in the fields.

Thus, before this young dog knew well what had happened, there sprang, as if by magic, from the snow a beauteous animal, strong of scent and fleet of foot, and heading straight away from him at top speed.

He heard a voice calling many names, and at the same time the crack of a whip. But his name was not among the rest; and he just had time to notice that the Over-Lord stood still, with the other dogs about him. Then he was off in pursuit, straight as a line for the river. There the hare made its first turn, Murphy being twenty yards in rear. He was running mute now, and both hare and dog were settling to their work—the one to escape if it could, the other to catch, if so it might be. They were through the far fence a moment later, and disappeared, only, however, quickly to return and take a line straight down this thirty-acre piece. It was a stretch of nearly a quarter of a mile, and ere they reached the further fence Murphy was gaining ground. The hare doubled at the boundary, and then doubled again, making the figure of a giant eight on the glittering golden surface of the snow.

Was the dog really gaining? It was a fine course. The hare was evidently a late leveret of the previous season; the dog was scarcely more than seven months old. How would it end? The Over-Lord stood and watched, determined that none should interfere. There should be fair play in a fair field, if he could only keep a grip upon these others that were whimpering and shivering and straining at the leash. He had passed the thong of his whip through the collar of the old dog, so all were really well within control.

Would the young dog last? That was the crucial question. The hare had had many a run before this to save her skin, and was hardened by the life of the breezy downs and the wide fields. But the dog had never previously been tried in such a way: his life had been more or less an artificial one, and he had never been called upon to lay himself out, or been put to such a strain as these almost maddening moments entailed. Catch this thing somehow he must. Were not his comrades looking on? Did not the very silence of the Over-Lord seem to demand of him his very best? There appeared, however, to be no getting level with this animal of surprising fleetness of foot, that seemed to glide over the ground with perfect ease, and that responded gamely to every effort that he made.

The group of lookers-on watched the more intently. Now the hare by a clever turn increased her lead; then once again the dog made good the ground lost. The hare had come back by this time almost to the starting-point. Closer and closer drew the dog: the hare seemed to be swaying in her stride. The dog's tongue was out at any length, and his pant was clearly audible. Once again the hare doubled, and the dogs with the Over-Lord gave tongue, as though they cheered their comrade. Then with a fling and a dash Murphy was into it: 26 there was a scuffle in the snow, and the next instant the young dog was seen to be holding the hare down.

Making his way to the two, taking the dogs upon leash and thong short by the head, and keeping them back by the free use of his feet, the Over-Lord seized the hare and rescued it; Murphy being too beat now to do more than lie stretched out, panting.

"Well, I'm...!" — The Over-Lord was passing a hand as well as he could over the frightened hare, holding it high to his chest. — "Run to a standstill, and not so much as harmed. Well, I'm...!"

He had let go the other dogs now. They were barking and jumping round him, and to avoid risk he was covering up the hare beneath his coat. His face was a study as he looked at Murphy lying in the snow. No fault was to be 27 found with the dog; that was very certain. He had been given an opportunity of showing what he could do. The snow had equalised the race. And this was the end — the hare not hurt at all. He would look again at her presently. It had been a pretty sight: Nature's working; no real cruelty in any of it. Such were the thoughts that were passing in the tall man's mind.

All turned homeward after that, the Over-Lord's feet scrunching the snow as he took great strides, a smile lighting up his face. Four of his dogs were close to his heels, as though they expected something; a yard or two behind followed a younger one, with his tongue out level with his chest.

Later on in the day, when all the dogs were kennelled up, the Over-Lord might 28 have been seen leaving the mill-yard, with something he carried in a bag, taking long draws at his pipe, and still with a smile upon his face. He was making his way alone to the open fields, and across these to where there was shelter under a hedge. Having reached his point, he stooped to the ground; and then there sped from him, as he rose, a hare, unharmed in wind and limb.

He looked long after it, to make sure. Then he rubbed his chin with his pipe in his hand, and remarked aloud, "Run to a standstill, and never harmed. Well, I'm...!" And once again that day he checked himself from using a bad, if sometimes almost pardonable, word.

29

III

The general company naturally viewed Murphy's performance from many standpoints. Among his contemporaries his reputation went up with a bound, though there was not wanting a leaven of jealous ones even amidst those who crowded most closely round him. Among those a little older than himself, the best-natured commended him outspokenly and in honest generosity of heart. Others, with more mundane outlook, judged his achievement reflected lustre on the kennel, and therefore — this with a sniff and the chuck of the chin — also on themselves. A few more vowed, in true sporting spirit, that they would do their level best to go one better if such a chance as that should come their way. To these last, the puzzle was why, with such results, the whole of those present had not tasted blood; and among themselves they voted the action of the Over-Lord incomprehensible, certainly womanly, very certainly misjudged. If the young dog had gone up therefore in their estimation, the Man had correspondingly gone down.

As for the older generation, some spoke patronisingly, as if they wished to convey that the deed was nothing more than they could easily have achieved, and in fact ended by talking so much that they persuaded themselves, to their own satisfaction, that they were in the habit in their younger days of doing things of the kind not less infrequently than once a week. The moralists wagged their heads as the fountain of all truths, and asserted that such success was a very bad thing for the young. The swaggerers, who held somewhat aloof, but who had never done anything in their lives, put on more side than usual and endeavoured to carry matters off that way, oblivious, as ever, of the laughter round the corner. Lastly, there was that other class, the crabbed and the crusty, who would, had they belonged to Us, have retired behind their papers in the Club windows, but as it was, and being dogs, merely made off out of earshot, with their ruffs up, grumbling to themselves and crabbing all things.

There were some of all classes here as elsewhere. It is indeed surprising how closely the dog family approximates to the human. The same counterparts are to be found in both. We mostly hunt in packs.

And if dogs are wont to bark and bite and rend, We, on our part, are often not behind in practising the same 32 strange arts, though not always with the same sportsmanship and generosity.

As for Murphy, he took the whole matter with a skip and a laugh, as if it was all part of the jolly fun of life, and as not in any way reflecting credit on himself. By nature he was modest and shy, and if he did things occasionally that were out of the common, he never seemed to grasp the fact, invariably looking puzzled and impatient at all praise. "Never mind all that; let's come on and look for something else," was what he said, exhibiting in this way, perhaps, one of those traits of character that made him so lovable, and that grew to such fair proportions as he advanced in years. His disposition was happy and generous, and though essentially manly—if such a term, without offence, is applicable to dogs—there was also about him a peculiar gentleness that was exemplified in all 33 his actions, right down to his inability to use his teeth. He was never known to fight; and, what was still more strange, bones were to him altogether negligible things.

For a character such as this to meet with harsh treatment, much less cruelty, was, if not to ruin it completely, at least to undermine all confidence. Yet this, sad to relate, was now precisely what befell. Up to this, life had been without a cloud. Of course, as in every other society, there had been the necessity of fending for oneself—of picking up a scrap, for instance, quickly, if you wanted it at all. Such things are good, and make for progress and development. But harshness and unkindness, like injustice, had been altogether foreign to the mill and all who lived or worked there. Life sped on in that favoured spot with as even a surface as that of the river, 34 whose waters flowed sluggishly up to the mill, barring the dam, and then went bubbling down the race, revivified and having done its spell, for the time.

How it came about is not now exactly discoverable; but just at this period of Murphy's life a decree was issued that several of the family were to be boarded out; and the next day the young dog found himself moved to the home of one of the mill-hands, half a mile and more away.

The cottage stood alone, and the family inhabiting it consisted of a man and his wife, and a daughter just finishing her schooling. Once there had been a son; but he, like many another in our villages, had gone out—all honour to them!—to strike a blow for his country some five or six years before, and had in quite a short while found a soldier's death. His photograph hung crookedly just above the 35 mantelpiece, with another of a group of his regiment by which he had once set much store, and yet another of the girl whom he had hoped some day to make his wife.

When the glow fell, and the bald, laconic message was delivered one winter evening at the door, the mother bent her head low; and later, when she found speech and had dropped the corner of her apron, was heard to whisper to herself, "'Twas the Almighty's will." Then the tears welled up afresh, as she rocked herself in her chair, gazing at the fire.

The effect upon the father was different. "What...!" he cried, as though some one had struck him. A single candle flickered on the table; his lips were drawn tight across his teeth; his fingers clutched the table-lid convulsively, and he leant across in the direction of his wife. 36

"What...!" he exclaimed again.

"They've killed un," repeated the wife, the candle-light reflected in her staring eyes. "Seth, Seth," she continued, following her husband, who had taken up his hat, and was making for the door—"oh, Seth, Seth—'tis the Almighty's will, man; I do know for sure it be;—Seth, Seth...!"

But Seth Moby had gone out into the night; and from that time forward he walked as one suffering some injustice. He had always been a man of uncertain temper, but this blow appeared to sour him. It is well to remember that once at least in his life he had loved deeply.

The Over-Lord brought Murphy to the door, and arranged matters with Martha Moby, just as he had often done with others in the same way. The day 37 had been wet; the lane on to which the garden-gate opened was muddy; the dog had dirty feet. "You'll take care of him, I know. He's a good dog—a good dog," he repeated, when he left.

It was after dark when Moby returned. "Wants for us to kep the dog, do 'e? There be a sight too many on 'em about; and for what he do want to kep such a lot o' such curs, nobody can't think. A-bringin' a' the dirt into our housen too. Err ... I'll warm yer!" he added, making as though he would fling something at the dog.

Murphy looked puzzled, and crept into a corner.

"Don't carry on like that, Seth; don't do it, man. The dog's a poor, nervous little thing with we, and don't mean to do no hurt."

But it was of no avail. Seth Moby looked upon Murphy as an interloper, 38 and when he could do anything to frighten him he did, and by any brutal means in his power. Even the mill-hands remarked to one another that their mate, Moby, was a changed man. "'Twas like that wi' some," they said. "Trouble sowered 'em, like, and made 'em seem as though they 'ould throw the Almighty o' one side. And once folk got on a downward grade, same as that, it wasn't often as they was found on the mending hand—no, it wasn't for sure."

On one occasion, after the first week was over, Murphy escaped, and appeared at the mill with a foot or more of rope trailing from his collar, for latterly he had been kept tied up. Seth chanced at that moment to be leaving work, and brought the dog up short by the head, by putting his foot upon the rope end almost before the dog knew that he was there. He half hanged him taking him back, 39 and flung him into the house with an oath that frightened his child, and made her run to the back kitchen that she might not hear what followed; while the dog crept on his stomach to the corner, his tail between his legs: he always moved in this way now, though it is said he never whimpered.

"Oh, Seth, if you goes on like this," said Mrs. Moby reproachfully, "there'll be murder, and then trouble to follow: the Master is not one to put up with cruelty to any dog. Bless the man—you're gettin' like a mad thing. Leave the dog alone, I tell yer." Seth had taken off his boots, and flung them at the dog before going up to bed: Mrs. Moby had been engaged trying to disconcert his aim.

That night another foot was heard on the stairs; there was whispering in the kitchen; and for several succeeding 40 weeks, and unknown to others, the dog slept happily with the child, though not without serious risks of trouble being thereby made for both.

At the end of that time the Over-Lord called. He had been away. He had heard on his return that all was not well with the dog, and had come to see for himself. Murphy had been lying curled up on a sack in his corner, but when he heard the well-known footstep he crawled out, hugging the wall nervously till he reached the door.

"Murphy, lad!" exclaimed the Over-Lord, looking intently at the dog—"Murphy, my little man; that you...!" The dog was fawning on him, saying as plain as speech, "Take me away with you; take me away."

The Over-Lord put his hand down and patted him. He did not say another word, as Murphy followed him out, save 41 "It's not you, Mrs. Moby; it's not you." He had a great heart for dogs, and began to blame himself on his way home for what had evidently occurred. "If the man did not want the dog," he muttered, "he had only got to say so; besides it was his rent to him: it was not done on the cheap—that never does in any line."

When he reached his own house, he took the young dog in with him—a thing almost unprecedented, so far as the rest of the outside company were able to recall. They judged their former companion spoilt, or on the high road to being so.

"It was all that hare," remarked the middle-aged.

"Yes," agreed the moralists—"success is always pernicious to the young!"

Lookers-on generally misjudge, though they claim to see most of the game. 42

The next morning, by strange coincidence, a letter was delivered at the mill, destined to alter Murphy's future altogether.

IV

Daniel was one of those dogs that die famous, though belonging to a small circle; not famous in the sense in which the dogs of history are so, but because he possessed individuality and stamped himself upon the memories of all who ever met him. And these last were not few, for Dan had travelled widely and had gathered multitudes of friends. Then, again, he possessed those two almost indispensable adjuncts of popularity—delightful manners and a beautiful face. It was his invariable custom to get up when any one came into a room; and when he advanced to meet them, it might certainly have been said that, in his case, the tail literally wagged the dog, for his hind-quarters 44 were moved from the middle of his back and went in rhythm with the tail. His looks were perfect. Being by Pagan I., he possessed not only eyes set in black and a coal-black snout, but also that further characteristic of dogs of his date, the blackest of black ears—a feature now entirely lost in the case of Irish terriers, and never, it is said, to be regained.

Apart from a liberal education and the miscellaneous knowledge he had picked up for himself, to say nothing of a wonderful series of clever tricks, the instinct known as the sense of direction was in his case developed to an altogether abnormal extent. Definite traces of this were noticeable when he was still a puppy; but it was at all times impossible for him to lose his way. As he grew older, this instinct became so marked, that it set others wondering whether or not there existed 45 among dogs a sixth, and perhaps a seventh, sense, lying far beyond the grasp of human, limited intelligence.

Dogs, as we all know, are not the only animals, that possess this mysterious instinct. They share it with many other classes, such as those of the feline tribe, and also with the birds and a number of insects. In fact, all animals appear to possess it in varying degree; they are all more or less able to find their way home. Yet, study it how we may, we are at fault when we try to account for it. In many cases, the homing instinct is apparently governed by sight; but many scientific observers entertain the idea that the sense of smell, in the majority of instances, will be found to lie at the root of the matter. Possibly they are right.

When, however, we are brought face to face with an exceptional exhibition of the sense, we have to confess that we are 46 left unconvinced by any of the theories that have at present been advanced. It is no unusual thing for a dog to find its way home along a road it had not previously travelled, going with the wind, and in the dark. One case is known to the writer where a dog found the ship it had come out in in a foreign port to which it had been taken, and made a voyage by sea, as well as a considerable journey by land on its return to this country, in order to reach its home. A cat also, within the writer's knowledge, found its way back to its home, though it had been brought some distance in a sack lying at the bottom of a farmer's gig, and though the return journey entailed traversing the streets of a busy town. Any one may test a bee's powers in the same way, by affixing to it a small particle of cotton-wool. When liberated, it will take a perfectly straight or bee line to its hive, 47 though this lie at a considerable distance. It is unnecessary to refer to the achievements of carrier-pigeons, when set free after a long journey and the lapse of many hours, or to the way in which rooks, especially, as well as starlings, will find their way to their usual roosting-places across wide valleys shrouded in dense November fogs.

Nor must we succumb here to the temptations offered by the very mention of migrants, though we may well ask, what is the power that enables a swallow to leave the banks of the Upper Nile and arrive at the nest it left the year before, beneath the eaves of a cottage standing on the banks of the Upper Thames? Or what directs the turtle-dove, year by year, from the oleander-grown banks of the streams of Morocco to the more grateful shade of our English woodlands? Yet marked birds have proved 48 the truth of these and still more wonderful achievements.

Instinct, the dire necessity of obtaining proper food, the perpetuation of the tribe—Nature's most imperious laws—lie no doubt at the back of many mysteries. Yet to say this is not to account for the sense before us, any more than it is to solve those innumerable problems that are scattered all along our several roads, and that we stumble over every step we take. Leaving out of count such systematic, and apparently scientific, labours as those of the ants, bees, and wasps, we constantly find in the animal kingdom powers being exercised, as, for instance, in the case of the earthworms and the moles, that are not to be explained by the use of the words instinct, intelligence, and necessity. The humblest of animals appears often to be handling forces with ease and familiarity, the range of which it must apparently, 49 if not obviously, be unaware. But if this last is true, and these animals that are blind walk blind, what are we to say of ourselves, when we are frequently doing the same, and handling forces that we are totally unable to define?

The digression is a lengthy one; but even now a further step must be taken. The man has, in the dog, his one real intimate in the whole animal world. It will be generally admitted that the dog depends exceptionally upon the man and the man often largely also upon the dog, and that in this we have yet another instance of that interdependence that is to be found throughout Nature and wheresoever we look. This, however, is not the chief point in considering the relationship existing between the two. There is something much deeper, and that goes much further.

Man, we are told, holds supreme dominion 50 on Earth. He is King over all things living, both great and small; and this constitutes at once his endowment and his responsibility. Yet this supreme power is being perpetually modified, not only by the forces he seeks to control—whose so-called laws he has to obey, if they are to be subjected to his use—but also by those very creatures to whom he stands in the relation of a King. It is here, in the animal kingdom, that the action of the dog once again stands first; for what powers of modification and influence can transcend those which effect a frequent and practical impression upon the actions of this so-called King,—by appealing, as the dog often does, to man's moral sense; by claiming love outside man's own circle, in return for love given without stint; by calling for a wider self-sacrifice, in the light of a trustfulness and loyalty that is exhibited here 51 and nowhere else in Nature in the same unfaltering degree?

The dog does all this and more, as will be shown, and by ways and instincts that are as unfathomable as the one to which reference has just been made.

It is time to return to the more homely matter of Dan, that instances may be given of how, on one occasion out of many, he exhibited the possession of the sense of direction, and also of the eye he had for country.

The writer had to make a journey to a neighbouring town by rail. The distance as the crow flies was not more than six miles, but the railway journey took the best part of an hour and entailed a change and waiting at a junction. Daniel accompanied him, having never made the journey before, or visited the junction, or the station of the town referred to. On arrival, the writer elected to 52 walk. Now Daniel was almost entirely strange to towns, and, though all went well at first, he finally succumbed to the fascinations of the streets, and disappeared. Every means were at once taken to find him; the police station was visited, the cab-drivers were warned, and a reward was offered. In the end, the writer had to return without the dog, and face the reproaches of the family. A gloom fell upon the house for the rest of the evening. But soon after ten o'clock a bark was heard, the front door was thrown open, and Daniel entered; in a state, it may be added, that bordered on hysterics, and with the tail wagging the dog more violently than ever. It was seven hours from the time he had been missed, and no light was ever thrown on how he had accomplished the journey.

A dog's memory is proverbial. There is ample reason for believing that many 53 dogs, when once they have smelt your hand, never forget you. But they also often appear to make mental notes of what they see, and to retain these in their minds. A retriever that has worked long on an estate will be found to know the position of almost every gate and stile in every field, and will use his knowledge instantly as occasions arise. He equally appears to know the rides of the woods within his beat, and where they lead. In other words, he has, in hunting parlance, an eye for country; and here is an instance from Daniel's life by way of illustration.

To reach a neighbouring village on one occasion, the writer used a tricycle. There was only one road to this village, distant five miles, and this was bounded on one side by woods and on the other by the river Thames, which it was necessary to cross at the outset. Here and there between 54 the road and the river were houses, the gardens and grounds of which were surrounded by walls and fencing extending to the river-banks. The tow-path was on the further side.

It chanced that after three miles had been traversed, another tricycle caught up the writer and passed him. Dan was ahead, mistook this machine for his own, and went on out of sight. The weather looking threatening, the writer decided to return home, feeling confident that the dog would discover his mistake and follow. A bicycle now overtook the writer, the rider of which, in answer to inquiries, said that he had seen an Irish terrier entering the village he had left, three miles back, cantering in front of a tricycle. There was nothing to be done but to go leisurely home, waiting every now and then to see if the dog was coming, while growing always more and more uneasy at his non-appearance. 55 At last the home was reached — and on the front-door mat sat Daniel!

The dog was perfectly dry, and had still the dust of the road on him. He could not therefore have swum the river; moreover, he had no taste for water. Equally, he had not come along the only road; while it was impossible for him to have travelled through the woods or along the land lying between the road and the river. There was only one solution of the difficulty, and this was undoubtedly correct. In his walks along the hills the dog must have noticed a railway in the valley and its bridge across the river. He had certainly never been along this railway or over this bridge. But he remembered its existence when he was lost, made his way to it, got over the river without the necessity of swimming, and reached home across country in time to meet his master, and with an expression 56 on his face of, "Well — what do you say to that?"

One more story of him must be given, showing his extraordinary sagacity as well as his determination. When he had set his mind on anything, brick walls were well-nigh powerless to stop him. He obeyed one man, if he were by; in his absence, he acted solely in furtherance of the plans he had in mind, and always with a knowing expression on his face.

He was paying a visit in the West of England, and had quickly found his way about. One day at luncheon some one was rash enough to remark in Dan's hearing that the carriage was going out. To run with the carriage was strictly forbidden, and this Dan never failed to resent, as he did also being shut up before the carriage came round. "Carriage" was one of the thirty-eight words with which he was intimately acquainted, and 57 when he heard it used on this occasion he may have made mental notes concerning plans to which he vowed he would be no party. However this may have been, shortly before the hour arrived for the carriage to start Dan could nowhere be found.

The road leading from the house branched into three at the end of about a mile; and, as this point opened to view on the afternoon in question, a yellow figure was seen to be standing there motionless, evidently waiting to see which of the three ways the carriage would take. Needless to say it was Dan, and that of course he had his run.

But an end must be made of chronicling the further remarkable achievements of this wholly remarkable dog — his sage comments as he grew older, his faithful discharge of his duties as he roamed the passages at night, his intense love of 58 sport and his deeds in that field in spite of his being hopelessly gun-shy, his large heart, and those beautiful manners which he still made pathetic efforts to show, even when he moved with great difficulty and was both deaf and almost blind. He was just a high-bred gentleman; and he had about him something of the courtesy of the old school, which will still be discernible in some dogs when we have finally and altogether lost the art ourselves.

Daniel was now growing old, if indeed he had not already done so. It was obvious that he could not last much longer — perhaps a year; not more — and it was necessary, therefore, to find an understudy. Irish terriers had been a part of the household for many years. Yet another must be discovered, though, as all agreed, there could never be another like Dan.

Thus it came about that inquiries were 59 made in likely quarters, and a letter was despatched to one who could be trusted, and who was known the country over for the dogs he owned.

60

V

"Yes," came the answer; "I think I have just the dog to suit you. With an old dog in the house such as you describe, every dog would not do; but the one I speak of is a *good* dog, with good manners and a very gentle disposition. You know that I do not make a practice of selling my dogs, but you shall have this one for – – guineas, and I will send him along any day that may suit you.

"I forgot to say he is well-bred; Postman-Barbara. He is entered as Murphy."

Two days later a dog's travelling-box was put out on to the platform of a little country station, and there and then duly opened by the writer. Lying at the bottom in some hay was a poor, cringing little animal, that had to be lifted out, and then lay flat upon the platform. In such terror was he that nothing would induce him to move; and the only way out of the difficulty was to take him up, while others smiled, and walk out of the station with him.

At a quiet turn of the road the dog was put down, being somewhat heavy, when once again he could not be persuaded to walk, or even to stand upon his feet. Again and again he acted in this way, till at length the house was reached and he was deposited on a mat by the fire, close to a bowl of good food.

And this poor little abject was Murphy! — Murphy, the dog with the pedigree of kings and even emperors; the dog that had run a hare to a standstill; the dog of the happiest disposition of any one in the kennel, and that had been the favourite and playmate of the whole great company. If this was what pedigrees were likely to produce, better to make a clean sweep of the hereditary principle at once; if this was a picture of a happy disposition, better to try what chronic depression had to show. A sorry favourite this. Up to now a suspicion had been entertained that a playmate should at least be gay. It was all evidently a mistake.

"Murphy!" — Why, this half-starved-looking thing that refused to stir or eat did not even know his name. If a move was made in his direction, he hugged the ground closer than before, shifting his chin backwards and forwards on the rug in abject terror. The coast had purposely been left clear, and Dan was out with the rest of the family.

Presently one looked in, and passed sentence without more ado: "Oh, you poor, miserable, shrunken little thing. We can't keep a dog like that—it is impossible!"

Later, Dan appeared. The young dog got up, went respectfully towards him, and licked him deliberately upon the lips. Dan wagged his tail. They were friends. Then once again the newcomer crept on his stomach to the corner of the hearthrug, and remained there cringing when any one went near. What did it all mean?

Nor were matters any better when the household retired for the night: in truth, they were much worse. The most mysterious sounds ascended from the lower floor, and grew steadily in volume. They woke one and then another, till at last they drew some one from her bed. Such unearthly groans had rarely before been heard from throat of living thing. Of course it was the "new dog," as he had already come to be called, for he surely was not worthy of a name.

A conference was held next day as to what could possibly be done, though with the usual result that some said one thing, some another, and nothing was definitely decided on. Had the matter been put to the vote, the dog would almost certainly have been forthwith returned from whence he came, in spite of a remark from one quarter that such a course might result in something serious.

"'Give a dog a bad name...' We all know the rest. To return this dog is for him almost certainly to be shot—at least, I wouldn't give a penny for his life."

Murphy meanwhile lay curled up tight on his corner of the hearthrug, with his eyes wide open, watching every movement intently. Dan said nothing, and went his way, voting the house to be upside down.

That day passed without improvement, though every effort was made and a walk was taken in the fields: the night, the stranger spent in company, for he appeared to have a dread of being left alone. The day following matters were unfortunately made worse. It is the fate of many who are down to find themselves trodden on: the lucky meet with luck; the unlucky, more often, with misfortune. The world is full of remarkably strange ordinances; or rather, it might be said, life is replete with incidents that are often the last wished for. From him that hath not shall be taken away, not alone that which he hath, but even that also which "he seemeth to have." So be it. No doubt, in the majority of instances, he deserves to be so made bereft. On some, however, such things come hard.

The room in which Murphy had taken up his abode was part library, part 66 studio, and part a good many other things. A large picture—the canvas measured six feet—was being worked upon on this second morning after the young dog's arrival; and, as was perversely ruled, it was just here that an accident occurred that might well have been judged impossible. The easel, in fact, with its huge canvas, was overset, carrying many things into limbo as they fell; and with the fate that too often pursues the unfortunate, Murphy therefore found himself suddenly buried beneath a mixed assortment of articles to which he had hitherto been strange. To add to the rest, a whole string of cattle and sheep bells, brought from various parts of the world, were set ringing, and others were dislodged; and for the moment it appeared that the dog must certainly have been killed. The only good thing subsequently gathered from the strange proceedings 67 was that the dog had uttered no whimper. But if he was frightened before, he was terror-stricken now; and matters had therefore gone from bad to worse.

There is little need to describe what followed. On the one hand, it was judged that this was the proverbial last straw; that the dog would assuredly never recover now; and that therefore the only thing to be done was to send him back, with an earnest appeal for his life to be spared. Yet, once again, cooler judgments in the end prevailed. The dog had not whimpered. There was something in that. Moreover, by what had now occurred, an injury had been done to his already unhappy spirit, and, unless all honour had ceased to find a place between man and dog, reparation was certainly his due. In one quarter a sense of pity had furthermore been generated—a 68 fact, though unsuspected at the time, that was to prove the hub round which Murphy's whole future was destined to revolve. An appeal to the heart, if such once gets home, can never really fail—unless, as Murphy's countrymen might say, the person appealed to proves heartless.

Thus it was that a sheet of paper that left the house the same evening contained words to this effect:

"I ought to have written to you before about Murphy, as also to have sent you the enclosed cheque. But, to tell you the truth, I have been so much puzzled by this dog that I have purposely waited a day or two before writing to you. I have owned dogs for a great many years and of many breeds and temperaments; but never, in the whole of my experience, have I come across any dog as nervous as this one: it is pitiful to see him. Even my old dog's presence does not help him; and really, so far, I have been able to make nothing of him. Perhaps he may get better; but I almost doubt it. I wonder if, without you knowing it yourself, the dog has been cruelly treated. I keep looking at him and wondering, for I cannot, somehow, link this dog lying in front of me, and never closing his eyes, with the description you wrote of him. The journey would not account for it. However, we must hope for the best."

To this came answer:

"In face of what you tell me of the dog, I cannot of course accept your cheque, and therefore return it. But do please keep the dog for a month or six weeks, or as long as you like, and write to me again then. I assure you the dog is a *good* dog. Perhaps his surroundings are strange to him. They must be. The old dog will help him to come round, I feel sure."

A few days later the door opened, and a stranger was announced. Murphy was on the hearthrug, as usual; the canvas and easel had been banished to a corner, and an effort was being made to accustom Murphy to the clicking of a typewriter—a sound concerning which he was evidently doubtful.

"Ah, Murphy; you're a nice dog, aren't you?" The dog had gone to the door, and the great figure of the Over-Lord was stooping to notice him. "I always like to see where my dogs go, if possible," he added; "and I wanted to hear from you, as well as to see for myself, what was the matter, for this is a good dog—a nice dog: I know he is. He'll come all right. Just please give him time; and then, if you don't like him, send him back. He is as good a dog—gentle, you know, gentle—as I've bred. Why, I can assure you, I refused (mentioning several hundred pounds)—I refused that sum for a pair of his relations, only last year; so you will judge he is well enough in the matter of class."

"Why did you refuse? Most people would have jumped at such an offer."

"Well—I'll tell you. I didn't like the man's face that wanted them; nothing else: I always like to see where my dogs go and the people they go to; and, after getting your letter, I determined to make the journey here, as soon as ever I could get the time. He's a nice dog; a good dog—I'm sure of it."

"You don't think there is anything in the suggestion I made to account for his extreme nervousness, do you?"

"Well—I know now that there is. I only got to the bottom of it, though, this morning. These things aren't arrived at in a minute, you know. One working-man very rarely splits upon another."

Then followed the whole story. "It was cruel—cruel," he jerked out at the end, finishing with, "I may as well tell you, I never liked the man. Latterly his work was anyhow—went from bad to worse, and I discharged him."

There was silence. Two great big men were sitting looking at the dog lying between them. The dog's eyebrows moved continually: his brilliant eyes travelled from one to the other; and presently he heaved a deep sigh, as much as to say, "It's all quite true—quite true."

If there had been hesitation about keeping Murphy before, there was an end to it now. Here was a dog—a young life—that had once, and not so long ago, been the delight of the kennel, the very embodiment of light-hearted fun and happiness; the most promising of all the 73 younger lot, and one that had never been guilty of wrong. Send him back! Give him up! What might his fate be if he went elsewhere? Death? Look at him. Look at his large brilliant eyes. They betoken nervousness, of course—inherent nervousness, probably. A cruel injustice had been done by this dumb thing, and by one of Us. Give him up! Clearly everything most prized was at stake, and claimed the exact opposite.

Why should a different justice be the lot of a dog to that meted out to a man? Is the superiority all one way? Each man knows in his heart that it is not; that the dog is often the better of the two.

How the thoughts raced through the brain!

"Murphy?" It was his new master that called him now.

Perhaps the presence of the Over-Lord had given the young dog confidence: *he*, at least, had been linked with happy times. Murphy got up hesitatingly and came to his new master's chair, with his ears drooping. He even suffered himself to be taken into this new master's lap, though not without great nervousness.

And after that the Over-Lord rose and said good-bye.

"No, Murphy, we won't part," were the last words he heard as he left the door; and this was the last time the generous Over-Lord was destined ever to set eyes on Murphy.

VI

Others laughed when they heard the final verdict, and called the undertaking hopeless and sentimental. The hopelessness remained to be proved; and, as to the sentimental part of the business, some one averred that sentiment lay at the bottom of most things. It might be unpractical from a philosophic point of view, as well as often fitting matter for a jibe; but sentiment, all the same, was generally a source of strength! Without it neither nation nor man would be likely to get far; it reflected the noblest part of man's nature, and touched a nation at its quick, if flags meant anything, and were to be followed and set store by.

There was quite a bandying of words over the matter. This dog was so different to Dan. It was not a matter of argument, certainly not on abstruse points. The dog had been broken in nerve, and admittedly by ill-usage. Probably he had been nervous from the first, and there was therefore all the less chance of his recovery.

To this was interposed the fact that many well-bred dogs are constitutionally nervous, and continue to be so all their lives, their condition in this respect being probably largely due to their brain development and increased powers of imagination.

That might be the case, came the answer; but all the same—how about the tail? The nervous organisation of this dog and his imagination had to do with his brain, which his eyes showed to be capable of development. These points had to do with the head. What about the other end? The index to a dog's character, as well as to his immediate proceedings, lies, as we all know, in his tail—the angle at which it is held, the way it moves or remains stiff and immovable; its position before a fight, its twist to one side when stalking, its confident carriage when the owner has "got his tail up." All these are so many signals, generally recognised by man and other dogs alike. Granting all this, what was to be said here? This dog had now been several days in the house, and no one had apparently seen his tail: it had been kept firmly down, and in such a way as to suggest that had it been long enough it would have been well between his legs.

At this, some one said that he had seen it once, and it was bushy; the only effect of this remark being to elicit the rejoinder that "*then* it wanted pulling." Another averred that, of course, nothing could be hoped for till he got his tail up: the job was how to set about securing so essential a condition in the case of the tail of this particular dog. No doubt the first thing to be done was to win him to the habit of standing on his feet: it was obviously impossible to attempt anything with the tail till this was achieved. So far, his attitude had been best describable as that of the prone position. If anybody moved, he crouched still lower; if he was persuaded to enter another room than the one he had particularly taken to, he grovelled; if there was any sudden movement or noise, he was terror-stricken; and, added to all this, it was obvious that he could never be a watch-dog, for he refused to sleep alone.

Of course he ought to have gone back; and all these notions about "bringing him round," giving him another chance and a happy life, were so much high faluting rubbish.

In the face of such arguments, based, as they obviously were, on universal testimony, even the faith of the person most nearly concerned and wholly responsible must, it was judged, eventually give way.

But if counsels and opinions alike failed to alter the decision that had been come to, they equally also supplied no answer to the momentous question—how, seeing he was to be kept, was the confidence of this dog to be won? There was hope in Dan, of course. He would teach him plenty of things, and tell him much besides. A good deal of faith was placed in this direction. But, even then, what about the general training? This dog would run riot, be disobedient and unruly, hunt when and where he should not, like other dogs before him, or even run sheep. If these things happened, what was to be done? To thrash him would be almost an act of cruelty by a dog of such a temperament: it might make him more nervous than ever, even if he could be caught for the purpose and made to understand the rudiments of cause and effect. Dan had learnt to "come and be thrashed," when such was necessary and he was summoned in those most ominous of words. It might be possible to teach Murphy in the same way: dogs, somehow or other, were almost universally capable of differentiating between justice and injustice, and bore no resentment. The reflection gave relief. Yet what would be the effect upon this dog if Dan was in trouble and took to shouting "Murder," as he usually did long before he felt the stick?

The problems were many, and grew in number the more the whole matter was considered. Two things shaped themselves from the first: there must be absolute fairness and justice; and, what was of no less importance, there must never be any trace of loss of temper in what had to be done, however trying the case might be. To show anger, to give an extra stroke when the stick was up, to be hasty for an instant, would be to fail ignominiously, to the mutual unhappiness of both.

The whole enterprise was thus obviously full of pitfalls. Yet faith declared this way: by kindness, sympathy, and self-control the end might be attained, confidence won back, the young life put into touch with happiness again.

As the further aspect of the question was considered, it looked rather as if, while the man was trying to train the dog, the dog might equally be all the time training the man. Here was one none too strong, whose nervous organisation had been shattered, and whose confidence had been wholly undermined. To win back what had been lost would be difficult enough in the case of a man; how would it be in the case of a dog? Oddly enough, too, the conditions of life of neither party here were of the normal kind—in one case never could be so. Yet here were these two, and by the merest chance, placed in juxtaposition. A strange link was forging itself apparently, quite unknown to both, and coupling the one firmly to the other, though neither was aware of it.

It was not until some time had passed that the position took a more definite form, and the question repeated itself—what if sympathy grew up and blossomed into something fair, with love and mutual confidence as its accompaniments? Such might result, perhaps. The thought added interest to the problem as it floated through the mind and was lost again.

There was nothing uncommon in the possible situation; it had occurred again and again. History furnished innumerable instances. Folklore, with its roots in truth, told endless stories of similar complexion. The Dog and the Man; the interdependence of both: living things of like passions — sharers of like passions; fellow-helpers, the advancement of the one having kept pace with that of the other, right up from the days when, in prehistoric times and the Neolithic age, as is shown by the bones that are found, the dog shared the home of the man and partook of his food — right up from the days when the Egyptians, though they dubbed him unclean, worshipped this animal, and, because of his fidelity and courage, gave him a place as one among three who were to share with them the joys of Paradise.

The same story is to be traced through all the ages. Even Ulysses could shed a tear for Argus, hiding the fact as well as he might from Eumæus; and Tristrem and Ysolde, in the legend, took Hodain to be their intimate companion, because he had once shared with them "the drink of might." So, too, the great Theron walked as the close companion of the Gothic king; and Cavall became the trusty servant and liegeman of King Arthur. The huge white hound Gorban sat ever at the side of the Welsh bard Ummad as he sang his songs; and the beautiful Bran was the friend for life of Fingal. Most men have heard of William the Silent's spaniel, who saved his master's life; and many may have seen the form of the dog, fashioned in white marble, lying at his master's feet on the well-known tomb at Delft. We have each read of Scott's Maida. And if some, perhaps, have made a pilgrimage to that long and narrow mound in the vale of Gwyant which, according to tradition, marks the resting-place of the immortal Gelert, others have read of the faithful Vigr who never again tasted food when he learnt that Olaf, his master, lay dead.

The stories are without end; and romance knows no limits when dealing with the subject. The lives of the Man and the Dog are found to be ever intertwined. Yet is there always this besides — the rift in the lute and the familiar refrain, that the life of the dog shall be short, and that Man shall go on his way with his head bent, till such time as he shall become rich once more in the love of a new-found friend — if that be always possible.

No man, it has been well said, can be deemed unhappy who possesses the love of a dog; and none are too poor to win it, as none are too high to rejoice and grow glad in it. The dog, at least, knows no difference of class or place in his attachments. To him his home is his home; his master, his master and friend, whether his lot be to follow the tramp on the road, or to walk behind a king to the tomb. And perhaps it may be due to the mystery lying at the back of this wonderful intimacy and connection, stretching far back into an altogether hidden past, that to strike another man's dog unjustly is equivalent to striking him; that to hurt a dog with intent is to earn the worst of characters and to stain one's kind; and that for a dog to be in trouble and claim aid is for him to claim also the man's heart — even, as has many a time occurred, the man's life — to the infinite glory of both.

Nor has it been only on man's side that 87 such deeds of heroism have been exhibited. The man, the woman, and the child have undoubtedly gone to the dog's help at the risk of their own lives on many an occasion; but so also has the dog risked his for the sake of the man — not from any moral claim, not because life is a precious thing and must be saved, not because of that power which impels, and whose chief gift is the sense of after-satisfaction that comes even to the most disinterested; such things lie necessarily beyond the reach of the dog mind. What the dog does is done for love, because of his faith, and because, unlike any other living animal, he thinks, in his unselfishness, more of his friend than he ever does about himself.

On the shores of a lake in Travancore, not far from the remote cantonment of Quillon, stands a monument to the memory of a dog. He was left to watch his 88 master's clothes while bathing. Presently he was seen to be doing everything in his power to attract attention, by barking and running excitedly backwards and forwards on the shore. An advancing ripple was then discerned on the smooth surface of the lake, and the next instant the meaning of this flashed home. A crocodile had got between the swimmer and the landing-place, and was coming out to seize his prey. Hope might well have been stricken dead in the face of such a situation. But the dog did not hesitate. Plunging into the water, he swam out to get between the horrid reptile and his master, and thus to head him off. It meant his own certain death; but the saving of his master's life. A moment later there was a violent agitation of the water, and the dog had disappeared for ever. Thus there stands to record his splendid action this well-known 89 monument, erected by his master in deepest gratitude, and that passers-by might learn of what a dog is capable.

The incident is not the only one of its kind, and may be left to speak for itself. But the influence of that one act has probably been world-wide; and it is because of the exhibition of such qualities that the moral power of the dog reaches to greater lengths than is generally supposed. There is indeed ample evidence for believing that the beauties often traceable in the character of the dog re-act unconsciously, and for infinite good, upon the roughest of our own kind—by claiming unselfishness from those who otherwise may lay claim to possessing little; by showing what love may be under stress and strain, hardship and rough fare; by the exhibition of patience and faithfulness; by those instincts that make the 90 most depraved of lookers-on pause and think, and ask the question sharply— "Whence that?"

In Kingsley's *Hypatia*, Raphael Ben Azra, his head filled with a false philosophy, is made again and again to act otherwise than he would by the mastiff Bran.

The "dog looks up in his face as only a dog can," and causes him to follow her and to retrace his steps against his will. There are her puppies. Is she to leave them to their fate? He tells her to choose between the ties of family and duty: it is a specious form of appeal. To her, duties begin with the family; the puppies cannot be left behind. Nor can she carry them herself. She takes Raphael by the skirt, after bringing the puppies to him one by one. He must carry them, she tells him; and once again he finds himself doing the opposite of what he would: 91 the puppies are transferred to his blanket, and he and his dog go forward together.

"After all," he says to himself, "these have as good a right to live as I have.... Forward! whither you will, old lady. The world is wide. You shall be my guide, tutor, queen of philosophy, for the sake of this mere common-sense of yours."

He tramps on after that, "trying to get the dog's lessons by heart." He catches himself asking the dog's advice, till he exclaims irritably, "Hang these brute instincts! They make one very hot."

At last, by the dog's means and the example of energy that she sets, he is instrumental in effecting the rescue of Victoria's father. Then, as the distracted girl throws herself at his feet, and calls him "her saviour and deliverer sent by God," even Ben Azra has to admit that the credit is not in reality his. "Not in the least, my child," he exclaims. "You must thank my teacher, the dog, not me."

The experiences of the philosopher in the novel are only those of many in real life. Man is not the only civilising agent in this world of many mysteries. And if we often exclaim, "Bother the dog!" we have still very frequently to follow where he leads, and often to our most definite enrichment in the end.

VII

It was four months before any improvement was discernible: it was a year before confidence could really be said to have grown at all. In some directions it never grew. For instance, of labouring men, gardeners, and the like, Murphy always remained shy. It was in no spirit of unforgivingness, for he was perfectly civil; neither did he owe them any grudge, grudges being forbidden usually by dog law and only entertained by the poorest characters of all. Thus he never became familiar, even with those he met daily: his memory was phenomenal, and by passing by on the other side he showed that his associations in this direction were unhappy.

It fell to this dog's lot to live a very quiet life and to be thrown with few — either dogs or men. His days were regulated by his master's doings, and these again were regulated, of necessity, by method. The weeks came, and ran their course, and did not vary very greatly one from the other. There was the daily round of work — almost incessant work, life being supportable that way and in no other. There was the break, half-way through the morning, of a run of a quarter of an hour, wet or shine. There was the walk across country in the afternoon, also totally irrespective of the weather. There was the turn at night under similar conditions. That was the dog's day in winter-time; perhaps also the man's. In spring and summer both lived under the sky, and regarded a house only as a place to sleep in. Habit is second nature. Interests were many, and in some directions ran parallel — sporting instincts, especially, being quite ineradicable. Life for both was thus exceeding happy; and life grew always happier with friendship: that is as it should be.

With those he met Murphy was genial, if shy. He grew to love the members of his little home circle; though three of the quartet ever averred that, in reality, he only loved one wholly and altogether, and clung to him in a way that others noticed — folk on the land always referring to them, the country over, as "Him and his dog."

Were they not always together? The shepherds on the downs recognised them at great distances, for shepherds see far. The shepherds' dogs knew them equally well, and they see furthest. The ploughmen in the hollows caught sight of them against the skyline in the waning winter day, when the team grew weary as they themselves — which last fact, too, made 96 these best of men shout with full lungs, "Please, will you tell us the time!" The man with the hand-drill sowing the spring seeds; the poorer folk, men and women with their buckets, stone-picking in the chill, autumnal weather; the stockmen as they drove the cattle home, or called them from the lush fields with the crack of a whip — spring time and harvest, all the seasons through; in wind and rain, in the great heat, in the snow and the blizzard, it was always the same. And thus, in this unenclosed country, where there were great woods, but where hedges were almost non-existent, the men of the land would look up and pass the remark to their mates, with a jerk of the head, "Ther's 'im an' 'is dog; see?"

Outside the home circle — though, to be sure, a dog is, or should always be considered, a part of the family — Murphy's passion was for Dan. He invariably got 97 up when Dan entered the room, and often licked him many times upon the lips: he paid him every kind of attention; bullied him to play when out of doors; woke him when he judged it was not fitting he should be asleep; and, in fact, made a young dog of him again for a time, though Dan was really old. He already owed Dan a good deal, for Dan had initiated him into many things concerning rabbits, rats, and the rest, that all self-respecting dogs should know. Thus the old dog being an inveterate sportsman, Murphy followed suit — and both were, at all risks, encouraged so to be.

As Murphy furnished and grew stronger he naturally became more handsome, till passers-by would turn and remark upon the pair — the old dog and the young, lying on the bank of the river, patiently, while some one did mysterious things with paints; or they were seen returning 98 together in the evening, sitting side by side in the stern of a boat. They were certainly a very uncommon pair.

Dan's character had been, of course, fully formed long ago, and a truly wonderful character it was, as has already been related. Murphy's was still in the making. If the whole of the first year was a period of difficulty, the first four months might well have staggered any one undertaking a self-imposed task of such a nature. The ideal aimed at was never suffered to be out of sight, but, like most ideals, it had a trick at times of receding almost beyond the range of hope. It was not that the dog was continually doing wrong. Perhaps it would have been better if he had been, for then there would have been something tangible. The difficulty consisted in conveying to the dog what he should not do, without frightening him, and without getting 99 cross and losing temper. To train a dog that takes his thrashing, shakes himself, lays his ears back, and prepares for the next, oblivious of consequences, is not beyond the wit of man, though possibly a gift. But what is to be done in the case of a dog that is terror-stricken, even if the voice is raised? The position forms as fine a period of probation in its way as any that wilful man could desire; and at that the matter may be left.

The philosopher tells us that we advance more surely by making mistakes than we do by lines more usually held to be right. Murphy took the former and apparently correct course, like others before him. The first real stride he made was thus in connection with an error, and it did him a world of good. It came about like this.

By way of preface—what can possibly be more irritating to a dog than sheep? 100 Master and dog were coming home together, and were persistently mobbed by a party of a dozen. Both agreed that if any real pluck lay at the back of the attentions so freely bestowed, the view entertained of the proceedings might be somewhat modified. But both were well aware that there was nothing of the kind; that the bold front was a sham, that inquisitiveness was the origin of it all, and that funk in reality filled every one of those dozen hearts, however much their owners hustled forward or lifted up their heads and stamped.

How long would Murphy stand such gross effrontery? That was the question of the moment. So far, he had followed close to heel, with his tail down—though it is fair to him to say that latterly he had come to carry it erect. Possibly the sheep approached closer than any dog of spirit could endure, or one frightened the 101 others and they began to run away. In a moment it was all over; the sheep had turned tail, and Murphy was after them, and had even found his voice.

The field was one of five-and-thirty acres, so there was plenty of room for him to turn them this way and that. To continue calling was, of course, useless. Time was better employed in taking a grip of the feelings and deciding on what was to be done. To make matters worse, the farmer himself was seen to be viewing the proceedings from a distant gateway. He would undoubtedly expect the law to be carried out, and dogs that ran sheep to be either broken to better ways or shot. It made no difference that the sheep were not his but "on tack" in his fields. What was the lot of these might be the lot of his another day. A thrashing was, therefore, now imperative. But how was this to be administered, when the only weapon was a shooting-stick, and the site was the middle of a large grass field? The best thing to do was to sit down, and be patient.

A part of the dog's education had already been that he was to stop when his master stopped, and when the latter sat or lay down he was to come in. He had already responded in a small way to this training, and now he dropped his games with the sheep, left them, and came slowly back. He guessed that something was about to happen by his master's solemn silence, and therefore approached with caution. It is never necessary in the case of ordinary offences and with ordinary dogs to be over severe with the stick — if a suitable one is handy, which it generally is not. A lecture and a shaking does as well, with a tap or two with a stick to show it is there. Provoking as the incident had been, this last is what Murphy duly received. The shooting-stick was much brandished in the air, and the dog called "Murder," long and loudly. The delinquent was evidently catching it, judged the farmer; and he waved his arm and disappeared.

That was gained, any way: what about the dog? He had learnt what the rattle of the shooting-stick meant. He had also learnt that sheep were to be suffered in their stupid, irritating ways, and not chased. For a short while he took the matter to heart, being always woefully depressed when he even thought he had done wrong. But he soon recovered, and showed contrition in the winning way he had now begun to acquire — by coming up shyly from behind, and endeavouring to reach the fingers of his master's hand.

The whole episode proved a success — from the man's point of view, at least; in the case of the dog and the sheep no doubt it was coloured. Murphy had certainly acquired confidence by what had happened, just as a boy may, when he gets his first fall out hunting, and finds himself less hurt than he fancied would be the case in turning a somersault. Added to this, there was also gain in the fact that from that day forward he was immaculate with sheep, as will be seen.

Though Murphy was quickly judged as one who had been "born good," and continued to be so regarded all his life, it is not to be supposed that he never transgressed, and thereby never incurred the punishment of a shaking. He was canine, as men are human; the two terms are equally synonymous with error, and faults, one way or the other, have to suffer correction. But in his case, the faults of which he was guilty were almost invariably confined to those of a petty and irritating description—exhibition of nervousness when there was no need, failure in the recognition of his name, lifelong inability to get out of the way of traffic on the roads, which made walks along roads very rare occurrences indeed, and many others of a like nature. Had it been otherwise, where would have been the training for both? On the one hand, there was always the ideal of enabling this dog to regain confidence in the human being, and making him the merry, happy fellow he had once been; on the other, there was the test as to whether this could be done without loss of hope in the face of repeated and almost continuous failure, and without the exhibition of irritability or loss of temper when provocations arose at first a score of times on every day.

Of his pluck there was never the slightest question. Again and again he would charge, for instance, into a quickset hedge when his nose told him a rat was there, and come out a mass of thorns, and with the rat fixed to his lip or cheek. He would then simply knock the rat off with a fore-paw without whimpering, and hold it down that some one else might come and kill it, for he seemed unable, or unwilling, to kill anything himself. Then, again, he habitually went straight up to the most savage of dogs—several times at the risk of his life, in the case of well-known fighters twice the size of himself—and by his manner or his charm invariably came away harmless.

He could never be made to understand—and it is the cause of shame now to realise the irritation that this caused on many an occasion—that all the dogs in the world, any more than other inhabitants of the world, are not necessarily our friends, or intend even to be friendly; and that dogs, like those about them, are 107 frequently in the habit of quarrelling and rending one another without regard to feelings, and with little of the spirit of give and take that life and a common lot might elsewhere be said to demand.

He was often told these things, but it, as with many of his kind, he looked as if he understood, he never really doubted to the end that other dogs were at least, and of necessity, his friends. He did not court their company. They often seemed to bore him, and more and more the older he grew; but he had a curious way of inviting some to his house, and it was no uncommon occurrence to find a strange dog lying in the morning in the hall that he had sometimes brought a long distance.

Of his hospitality in this way he once gave a remarkable instance. A neighbour's dog was of uncertain manners, to dogs and men alike. One evening he 108 came to call. Now Murphy's dinner was always placed at six o'clock in one corner of the hall, and had just been brought when this visitor appeared. Not to be outdone in hospitality, Murphy at once pointed out the repast that had been spread, and stood by while the other ate, though he had himself had nothing since the early morning, and could, had he been so minded, have knocked the stranger into the proverbial cocked hat. All he did was to wag his tail and look pleased, as his dinner slowly disappeared. But, after all, such episodes as these belong to a later period, when he had become well-nigh human; when—it may as well be now confessed—he came to love the company of a man more than the company of dogs, when confidence had been won back, and happiness—happiness that with those he knew and loved showed itself in an intense and merry joy of life—had been finally regained. 109

One other peculiarity about him, or, rather, accomplishment, he possessed, must be noticed here, for, with a lifetime's experience of dogs, no parallel can be recalled, or has been gatherable elsewhere. First of all, he was certainly musical, and often after a long day's work, when the landscape outside was wintry, dreary, and wet, and the piano was thrown open and thrashed for joy of sound and relief,

Murphy would rise from his mat and come and lie close to his master's feet. He did not sing or howl on these occasions, in the way that with many dogs conveys the impression that music is pain. On the contrary, he remained quite silent, contenting himself with a sigh and a lick of the lips, which almost gave the impression that he would have said, if he could, "Just play that again, will you?"

This is, however, by the way. What he excelled in was what is generally 110 known as talking. The sound was not a howl, or like one; it came from deep in his throat, and was deep in tone, inflections being produced by movements of the jaw at the same time. To ask him a question was generally to get an answer in this way, though rarely out of doors, where his attention was necessarily distracted. But when once he had started, he continued to respond, and so to carry on quite a lengthy conversation. That was his sole trick, if indeed it could be so classed, for he evolved it entirely himself. Of tricks proper he knew none, and through life entirely declined to learn any. Perhaps Dan, whose repertory was large, had told him what a bore they were, and cautioned him to do his utmost to avoid them.

VIII

About a year after Murphy's arrival, Dan was gathered to his forefathers, and there was mourning throughout the house for many days. To one at least, if not to more, Alphonse Karr's remark held good—*On n'a dans la vie qu'un chien*—and Dan was that dog. His life had been long; he had won all hearts; he had done many wonderful things, besides fulfilling his duties as a faithful constable of the place in which his lot was cast; and now, loving and beloved, he had died. Such were the data from which his epitaph had to be evolved. Man could desire no better. To have been loved—that, all said and done, is the great thing, for it comprises all others. Another French writer reckoned it the highest eulogy bestowable, and it seems as if he was not far wrong, whether we have before us dogs or men.

One of Murphy's last acts by his grandfather reflected his own character, no less than the affectionate relations existing between himself and Dan. It was the custom to give the dogs certain biscuits after dinner of which they were particularly fond, and they sat side by side to receive them. One evening, when the biscuit tin was taken out as usual, Dan was absent. He was old; probably asleep: better let Murphy have his, and have done with it. The young dog refused to have anything to say to such suggestions; and for the moment his attitude was put down to an access of shyness, for these particular biscuits were irresistible. Presently he began barking and running backwards and forwards to the door. Being let through, he ran to another, found a third open, and presently returned in a perfect ecstasy of delight, with the old dog by his side. He subsequently referred to the extraordinary stupidity that had been evinced in a long and comprehensive speech. To steal a march on the old, or to fail to treat them at all times with respect, was evidently, in his opinion, wicked. At least, that was his text.

Dan's last resting-place was, of course, in the dogs' burial-ground in the family home. To lie there was the highest honour bestowable, and Dan had wholly earned it. Many generations of dogs lay in and around that corner, and the spot, if not consecrated, was at least regarded by most as very sacred.

This was it. An angle of old, ruined brick wall, facing West—part of an ancient garden—beautiful in colour and 114 overgrown with ivy. Great trees all about it; and the wide stretches of a park, where rabbits played in the long evenings, extending from it on all sides. A holly hedge and ha-ha prevented trespass; but those invited there found in this quiet sun-trap many headstones, bearing names and dates and epitaphs. Close by, a path, along which members of the family went often to and fro, led to yet another quiet corner, where a well-known spire showed above the trees. From this last there sounded at intervals the music of bells, chiming, or ringing solemnly, and beneath its shadow slept other folk, who had once walked the world with these same dogs of many generations, earning epitaphs no better, if as good, as they. To lie in either, seeing what falls to some, might well be thought a stroke of luck for dog or man.

It was not always so for dogs, here or 115 elsewhere, whatever it may have been for men. Within the recollection of all past middle age, dogs were kept tied to kennels by heavy chains, seldom allowed in house, fed at uncertain hours, and taken out at hours still more uncertain—if at all. Left often to howl time away by day, and to bark themselves to sleep at night. And when all was over, life having been often shortened by disease, there came along the man with the spade, detailed for the job, to fulfil the last of offices, and put in some handy resting-place the dog that had had his day.

We have come out of all that now, and rather plume ourselves upon the fact. We have altered our opinions respecting the proper place and surroundings of our dogs here; and many of us are not ashamed to confess that we hold opinions staunchly regarding their place and surroundings hereafter. We also have our 116 dog-doctors, our dogs' infirmaries, our homes and charities, and, in the end, our dogs' secluded cemeteries. Such things, in the case of dumb animals, point, we judge, to a higher grade of civilisation, and to many other things besides.

Yet let us not forget the fact that others, in the past, have gone before us, and far ahead of us, on this same track, of which we often speak with so much unction. In ancient Egypt dogs had names, and these are found inscribed in many places. They were the favourites of the home, and constantly made much of. They wore collars, too, and often by no means cheap ones; and just as they were everywhere admitted to the house, so, all these ages ago, they were talked to, and also made to talk. Legends were woven about their doings and their ways. And if, in many cases, they were small and insignificant, with short legs like the 117 Dachs, or, perhaps, the Aberdeen, implicit trust was placed in their fidelity as guardians of the home and family. Of course there were bigger fellows to fulfil the heavier duties, like the huge Kitmer, the dog of the Seven Sleepers, whom God allowed once to speak, and to answer for himself and others for all time. "I love those," he said—"I love those who are dear unto God: go to sleep, therefore, and I will guard you."

That was sufficient, surely. Then, too, there was Anubis, who was given a dog's head and a man's body: he was worshipped as a deity and the genius of the Nile, who had ordered the rising of the great river at the proper season from the beginning of the world, and whose doings in this way were marked by the coming of the Dogstar, with seventy times more power than the sun—the 118 brightest of all in the purple dome of the night.

An animal such as the dog, even if dumb, which in justice he could scarcely be thought, was thus judged entitled to a consideration never vouchsafed to others, and duly received it, therefore, at all times in this enlightened land. And not only in the fleeting years of his existence, but equally when he lay down under the common hand of death. The dog, in those forgotten days, received embalmment, just as his master and mistress, and was then carried with some solemnity to the burial-ground that was set apart for dogs in every town. And when the last good-bye had been said, the family to which he had belonged returned again to their house, and put on mourning for their friend and faithful guardian, shaving their heads, and abstaining for a time from food. So was it 119 with dogs all those thousands of years ago. We have not come so very far since then.

Murphy was not told many of these latter things, though obscurantism is always to be utterly condemned. It was thought better that he should not know them, or other darker facts to do with modern scientific times, lest by chance they give rise to strange and unorthodox reflections in a brain so active as his.

When the day came for Dan's best friend—she called him "Best of all"—to set out on a journey, to see the last of him, Murphy and his master, being left alone, turned naturally in their talk to the place where Dan was to be laid, as also to the doings of many other dogs who had lived and loved and had had the supreme happiness of hunting there throughout their lives. Some were good, and others, well not so good. Others were not thought much to look at, though this generally resolved itself into a matter of opinion. To set against these last, some were the very finest of their kind, such as Ben, the great Newfoundland, who had the glory of being painted in company with two small members of the family sixty or seventy years ago.

Each, of course, had his characteristics, and did his funny, or his wicked, things. In the face of a recent occurrence, it would have been a mistake to point a moral, or reference might have been made to Bruce, the deerhound, shot dead by accident when hunting sheep at night. That would do for another day, should circumstances arise to give the story point. There were plenty of other anecdotes besides that, and here are one or two that Murphy heard.

Perhaps Fritz, the Spitz, did the most remarkable thing of all. His master was an undergraduate of Christ Church at the time, and had been always in the habit of taking him with him on his return to Oxford. On a certain occasion he decided that Fritz, for once, should remain at home. The next day the dog was missing. Then a letter came, and this is what Fritz had done. He had found his way into the neighbouring town, distant three miles, and taken the train to Swindon, as was duly proved. Probably he changed there, though this is not recorded. But he went on to Didcot, where he certainly got out, found the Oxford train, and that same afternoon walked into his master's rooms at Christ Church.

One other action of his deserves to be recorded, for it affords an instance of how nearly dogs approach at times to human beings. No man is so wholly hardened as to care to die disliked, while many have a fancy ere the end to seek forgiveness, that they themselves may die forgiven. So was it with Fritz. Like many men of genius, his temper was uncertain, and on more than one occasion he was known to bite. The day before he died, though old and infirm, he made a round on his own account and visited one or two to whom he had certainly behaved badly. His action was recalled when once again he disappeared. But it was further remarked upon — some adding that they thought they understood — when Fritz was found curled in a hole beneath a bush — and dead.

Graf, another of the same breed, but belonging to a period twenty years later than Fritz, had also curious ways of his own. He could run down a rabbit in the open, and did it on many an occasion; but if this was remarkable — a rabbit being reckoned one of the quickest of all animals 123 for a hundred yards — his curious behaviour exhibited itself in quite another way. He was a dog of great character and cleverness, as well as perfect manners. It was the custom in the family at that date to have prayers on Sunday evenings. This Graf never failed to resent. There had been service in the church during the day, and Sundays were dull days for dogs: why have prayers in the evenings to make things worse? Therefore, to show what he felt in the matter, no sooner had the family left the room for prayers, than he gathered up the newspapers and tore them deliberately to pieces. It was not only once or twice or even six times that he did this. He did it repeatedly; and when the family returned, *The Guardian* especially was found in scraps upon the floor.

But he was otherwise a good dog, and so it was that he who read *The Guardian* 124 week by week on Sunday evenings showed that he bore Graf no resentment, for when the dog died he wrote a poem running thus, the last line and a half of which are graven on Graf's stone:

"Can such fidelity be all for naught?
Is virtue less true virtue that it beats
In a hound's faithful breast? No, Graf, the thought
Of thy pure, true and faultless life defeats

All doubt. No! Virtue lives for ever, and the same,
Whether in man, or in his faithful friend
Who looked but could not speak his love. The flame
That warmed thy faithful heart can never end

> In dark oblivion. If not a Soul
> Is thine, at least is Life. The same great hand
> Made thee and us; but where upon the scroll,
> At day of Judgment, shall be found to stand
>
> A human soul so faithful to the end,
> So true as thou hast been? God's great design
> Awaits both thee and us. Good-bye, sweet friend,
> And may our lives be simply true as thine."

By way of parodying this, in the case of another dog, it was suggested by one who was flippant that his epitaph might run — "And may our lives have fewer faults than thine." But while it is true that this one had run up quite a heavy bill in cats and committed many other enormities, the line *De mortuis nil nisi bonum* was kept in view, and, if nothing could be said, it was judged better to say nothing. Moreover, as Murphy duly remarked, while we talked over the wonderful doings of many and many a dog now lying in this sacred corner, "What could you possibly have expected in such a case, and from one of Us that you had wilfully named Scamp?"

There was, of course, something in that, and many of Scamp's acts deserved to be recorded, though this is no place for doing so. At one time he was in London. Residence there naturally put a limit to the exercise of his sporting instincts, but he developed others to replace them. He was sometimes absent all day, to be found at the door at night; and on one occasion he met his master at a City railway station, when thought to have been lost for good and all — was indeed seen by his master to be making his way thither as he drove into the station yard in question.

To have done anything so clever as that might have been thought to have earned the right to headstone and epitaph in full. Yet his resting-place remains unmarked, and his name apparently dogged him to the end, and past it.

"What was that about *De mortuis*?" came the question from Murphy.

"*Nil nisi bonum.*"

"That never should have been raised, in his case. What about *De vivis*?" There was indignation in the tone; perhaps justly.

IX

"What I does is this—what I does is, I gets 'em quite close to me, and then I talks to 'em."

This is what Mrs. Pinnix invariably replied, when asked how it was that her children were of such good behaviour and gave so little trouble. And Mrs. Pinnix knew, for she had been the careful mother of thirteen, and had developed this happy, good-natured method of dealing with each in turn, boys and girls alike. No doubt she was a remarkable woman in many ways, for she won the last event on the card at the time of the Jubilee sports, being then the mother of ten—"Skipping: open to mothers only." But the point here, in this remark of hers, is that a long experience with dogs shows 128 the talking treatment to be as applicable to them as it was to Mrs. Pinnix's children.

Nor will this be found to be the fanciful idea of the few, if inquiry be made. To live largely, for instance, among those whose labours lie far from cities, and who, of long habit, have come to note many things concerning which the less fortunate townsman knows nothing, is to learn many things oneself. To hazard the remark in such quarters, that a good many people have no belief in the theory that talking to a dog does him good, is to receive for answer, "Ah, but I knows as it does." Others go further, and in reply to the question whether they think dogs—that is, the best dogs—really understand what is said to them, never fail to assert with emphasis, "Well, they does; I be sure as they does: 'tisn't a mossel o' use to tell folks the like o' we different." 129 Shepherds, stockmen, farm labourers, old villagers who have had many experiences though living in a narrow circle, and who look back over a long life, constantly make use of such remarks. And probably dog-lovers of all classes will re-echo the same.

It was certainly the method adopted in the further training and education of Murphy. As already related, he had been taught to stop when his master stopped, and to come in when he sat or lay down. Thus, though he was generally allowed to range at will over the open lands and be sometimes far distant, in the event of the one he spent his life with lying down to rest for a while, very few minutes would elapse ere the dog would be found making use of shoulder, back, or arm as comfortable things to rest against. Tucked closely in in this way, his face was level with that other's, as, with ears cocked and those human eyes of his, he took stock of everything passing in the valley, or that moved on the edges of the great woods clothing the hill-tops.

That was the time to get hold of him; to train him not to run a hare that might come lolloping stupidly along, down wind, into the very jaws of danger; to take no notice of a rabbit that offered insult by drumming with his hind legs on the ground only a few yards off; to tell him strange stories of what he might expect in the years to come when he grew as old as his master, and had learnt to try to take many knocks, to face many problems, to bear and suffer much that might come from strange quarters—had learnt also how to live, and to reap his share of the happiness that the mere fact of living rarely fails to give to all who are not weak-kneed or chicken-hearted.

Of course experience, in some ways, tended to undermine confidence. Did he not know all about that himself? Had he not at one time come to doubt all things human? Had not happiness and trust and faith gone by the board, because of the hardness and injustice meted out to him? But what now? By some miraculous process there had come a change. Doubt had not altogether vanished; confidence had not altogether returned; faith and trust in the giants that stalked over the world, and who seemed to rule it, were not as yet quite re-established: perhaps they never could, or would be. To some natures recovery in such directions is impossible. The fire has seared, the cicatrice remains—though to be hidden away, of course. To show feelings—above all, to show you are hurt—to sing out, in fact—is to exhibit a poor spirit, to fall short in proper doggedness. Suffer in silence, if you can—that must be the rule; just as this dog, with his keen, eager face, loves in silence—loves all the more deeply, perchance, because he loves in silence, and because that silence is so much more eloquent than words.

Did Murphy understand? According to Job Nutt, the shepherd, who was a philosopher in his way, "of course he did—he know'd he did: his'n did; for why not your'n?" In the face of such definite assertion there was no room for doubt.

Nutt had had his lambing-pens, that year, down in the hollow where there was "burra" from the winds. It was snowing when the hurdles and the straw were carted out, and all hands had set to work building the sides of the great square, with their thick, straw walls, their straw roofs, the snug divisions into which the sides were divided, the whole sloping to the south to catch what might be of the pale, wintry sun. Every one knew that sheep lambed quicker and earlier when the snow fell. There had been no time to lose therefore. The first lambs would be heard a fortnight before Christmas. And, as a matter of fact, by mid January, Job Nutt's family already numbered sixty-three. That was of course nothing. Why, one January, his father had had one hundred and fifty-one lambs born between a Saturday morning at light and Monday, no fewer than forty-two being doubles—and snow falling all the time. Ay, and when he moved his hurdles—that is, those that were straw-wattled—they were caked so hard with snow that they stood upright of themselves. His father "had had to work *some* that day and them two night." And Job always grinned a merry grin when he told the story.

But now, to-day, when the two who were always together dropped down from the hill to pay a visit to this shepherd, it was the last week of February, when the mornings are as brilliant and full of hope as any in the year. The rooks were busy building in the great elms by the river; the wattles just below the lambing-pens were already turning red. Spring was coming: the colour of the sky, the voices of the larks, the bleat of the lambs, all told the same story. Of course winter would return: it always did. But, for the moment, there was a passing exhibition of beauties in store, a reflection of things that should be. By the afternoon the grey blinds would be down again. But that did not matter in the least: this glimpse had been permitted, and in the brilliant sunlight and the stillness the happiness of full confidence had welled up, and seemed to fill the whole world.

Murphy certainly appeared to feel it. As he and his master sunk the hill, he stretched himself out as he ran; he jumped into the air for joy. His doings, in some mysterious way, frequently reflected the colour of the day; and his spirits varied with those of his master. The sympathy of dogs is no modern discovery, but as old as their comradeship with man; and thus this one varied his ways according as times were good or bad, or trials, mental or bodily, chanced to be the same. On this brilliant morning man and dog had caught the light of the sun and the gladness thereof, and the young dog played with his master's hand as he swung along, and barked and jumped for very love of life.

He was often like this now when they were alone together, though, with others, he would sometimes lapse again into uncertainty and hesitation. Nevertheless, there was no longer doubt that he was on the right road: happiness had in a large measure returned; confidence was following. The man and the dog were drawing very close to one another, and in more ways than one.

The pens were only tenanted now by some thirty ewes, still to lamb, and by those "in hospital," as Job spoke of them. Four hundred tegs, ewes, and lambs were in fold on the hill, on a clover stubble, or what remained of it, being given crushed swedes and other things, for keep was scarce so early in the year. The shepherd's boy and his dog were up there with them: only Job and Scot were in the pens. Murphy knew this last, savage though he was; and had duly delivered to him, on many a previous occasion, that strange message of his that compelled the most savage to let him pass free.

"Oh! he can come: I likes that dog o' your'n," called Job, ordering Scot to his place beneath the bleached and weather-worn hut on wheels, in which all the miscellaneous articles of a shepherd's craft lay stored. "I be just about to find that mother yonder a new child," he added, with his usual grin. He was busy tying the skin of a dead lamb on to the back of another—dressing him up, in fact, in another suit, even as Rebecca once did Jacob.

"When a yo do lose her lamb, we's careful to leave the dead un next its mother, for they've got hearts same as we. If us was to go for to take the lamb, they 'ould pine. 'Tis nat'ral, ain't it? Well, you see, 'tis like this. After a bit we takes a lamb from a yo as has a double, like this un here; skins the dead lamb; and ties the skin round t'other's neck, same as this—see? She'll let this un suck then; but she 'ouldn't afore—no fear! They do know their own childern, same as we; just as they knows them as tends 'em. By-and-by I'll cut this skin away, bit by bit, when I judges this un has got to smell same as her own child: it'll be all right then. Ah! 'tis like this with sheep—there's something to be learnt about they every time in the day as one comes nigh 'em."

So the two men rested against the hurdles in the sun, and Murphy sat solemnly between them: he had become very particular in his manners when with sheep. The disguised lamb was already sucking the ewe; and Job lit his short clay pipe and smiled: he had been up all night.

"I'd never have a lamb killed, if it was my way; no'r I wouldn't. Do you minds last season, when you and yer dog was along? I wus a-going across the Dene with a bottle o' warm milk, with a bit of a tube stuck in it, if you minds. 'Twas warm milk I'd taken from the cow. Ah, well, 'twas for a lamb as had lost its mother: udder wrong; I could find of it when the master brought the lot in. And I goes for to say as any un as 'ud serve a yo that way should be crucified. Well, 'tis that very lamb as was as is now the yo a-suckling the one we dressed up. See how things do work round, don't 'em?"

But the talk was not always about sheep, when the folds or the pens were visited, or "Him and his dog" walked with Nutt and other shepherds over the open lands, in the wind and the weather.

One day Job had been busy sheepwashing, and the talk turned on dogs, as it often did.

"'Tis wonderful what they knows. What don't 'em know? I says. See that Scot I had—the one afore this un. Well, I was down a-sheepwashing, same as I've been just. One o' the full-mouthed sheep as we had then broke away, and went straight over river, and it ain't very narrow there, as you minds. She got up on the further bank and stud. And Scot, he looks at me, and across at the sheep, and then at me again. I know'd, right enough, what he wanted. He wanted to go over and fetch that sheep back. But I 'ouldn't let un, for a bit. And he kept a-looking and a-looking, same as any one might speak. So I just moved my head, like; there was no call to do no more. And off he set in the water, and swam river, ketched the sheep by the throat—oh, no, he didn't hurt un, no fear!—dragged un to the bank, and brought un over, right enough: he did, though."

"Well, 'twas like this," he continued, after a laugh. "A gen'leman was a-rowing by in a boat at the time. And he comes across to our side, when he sees what Scot 'a' done, and he says, 'Shepherd,' 141 he says, 'I'll have that dog off you, if you've a mind.' And with that he puts three golden sovereigns on the bank at my feet, where we was busy a-sheepwashing. So I looks at the sovereigns, and then at he, and says to un, with a laugh—I says, '*No Sir.*' Lord, how he did pray me to let un have that dog!

"Then it come about this way. That evening we was a-coming down through the village, and passed 'The Crown'—that was, Scot and me—and there stood the same gen'leman at the door. So he comes across the road, seeing me, and he says, 'Well, shepherd,' he says, 'will you part with the dog now, for, if so be as you will, I'll make it five instead of three?' he says. And that's truth. And I just looked he between the eyes, like, and says, 'Part with my dog, Sir?' I says. 'Why, Sir, if I wus to part with he, I'll tell ye what he'd do—he'd pine and die—he'd 142 just pine away and die.' And with that I passed on, and left un. Dogs—well, sheep, if you do please to understand, is sheep; but dogs is dogs, and God Almighty do know as they be wonderful."

"It's not all dogs, though, that are as shepherds' dogs, Nutt—or capable of being."

Nutt shook his head. The two men and their dogs were on the hillside, with two hundred and fifty tegs moving before them. The sheep were walking with a wide front, but in single files, following those parallel tracks that had marked this steep hillside for centuries, to puzzle strangers.

"You can't make a shepherd's dog out of every dog, can you?"

"Perhaps not, in your meaning. But I do know I could train a'most any dog, if as I'd be so minded." 143

Scot was on ahead, where he should be. Murphy was close to heel.

"Do you mean to say you could train this one to fold sheep?"

Job Nutt took a deep draw at his pipe, and turned and looked down at Murphy, now just over three years old.

"I likes that dog; well, I've allus liked un. Train un to sheep? I believe as I could, were I to be so minded: I do believe as I could."

The two had to part then. It was dusk, and looked like wet; moreover, some wether sheep in the fold, far down in the valley, were "howling" for rain: they were true weather-prophets always.

So he might be trained to sheep. Job Nutt's words kept repeating themselves in the mind—"I believe as I could; I do believe as I could." What the shepherd had said was a testimony to this dog's marvellous intelligence; but then every 144 one had come to testify to that and to remark upon it. He was of course nervous and shy, and no doubt would always be so. Perhaps it was these characteristics that gave him the further one of extraordinary gentleness, that won all hearts. Many had already said, with a laugh, that he was "born good"; but latterly some had come to add that he was incapable of harm or ill.

And yet with these characteristics, amounting as they did to a certain softness, there was never any question of his pluck and spirit. Nor was there any limit to it. He had the spirit and "go" of any dozen of his countrymen: what more could possibly be said? At the same time he had the gentleness of a child. He recalled to mind one of those characters that some of us have met, and in strange situations—situations and hours when men's spirits were on fire, and 145 when the air was filled with sounds that once to hear is never to forget. One such is recalled by memory now—a vision of a lithe and active figure that had come its longest marches, and borne the many hardships of the many nights and days, though looking frail as a girl in her teens, and with manner always gentle as a child. For one like that to be amidst such doings as these seemed incongruous. Yet had the estimate proved in the end quite false. Breeding and pluck—nervous energy—had carried through, when others had gone down. And the pluck and the breeding showed itself still, when the blood dripped, and ebbed away, and the face was white as a stone.

Nor is such a parallel as far fetched as might at first appear. Given the two, the dog and the man, this dog was to show before the end characteristics equally striking and of scarcely less charm. To 146 bear pain is not easy. There is no longer doubt that men feel pain in varying degrees, and that sufferings that might be considered identical are multiplied tenfold in the case of a highly developed organisation. With the high intelligence and nervous development of this dog, it might have been thought that pain would terrify. If so, he never showed it.

It is unnecessary here to refer to the many instances when his dash and high spirit brought about an accident, for all our dogs get into trouble and meet with accidents at times — at least, those of any worth. But it was this dog's further habit to avoid, when in pain, the company of the one he loved best, and to go invariably to a woman for aid. It was as much as to say that he knew that many men were in such cases worse than useless: a thrust in this instance not without its truth. Thus he came home two miles 147 one night in snow, with both fore-feet cut right across with glass — due to a dash at a rat in some rushes on the frozen riverbank. To his master's eternal shame he never found it out. But, on arriving home, this dog went straight off for attention, of his own accord, and bore what he had to bear, not only without a flinch, but showing his gratitude by licking the hand that was tending him. So again, when he was once badly stubbed, he went to the same quarter, showed his foot, and then lay down, staying perfectly quiet while a spike was looked for, at last found, and then pulled out with a pair of iron pincers.

These are trivialities, no doubt; but they would not be trivialities to some of Us. It is by such that character shows itself — is moulded and made up — for others to estimate and take due note of. And thus it is that whether they are exhibited 148 by man or animal, we admit their charm and pay our tribute to them, just as Theron's faithfulness to Roderick drew these words from the lips of the aged Severian:

"Hast thou some charm, which draws about thee thus
The hearts of all our house — even to the beast
That lacks discourse of reason, but too oft,
With uncorrupted feeling and dumb faith,

Puts lordly man to shame?"

149

X

The hay harvest had been a light one, owing to the weather in the spring and the absence of wet. It was hardly off the ground before the corn harvest had begun and the long arms of the self-binder were to be seen waving in the air above the standing oats, the first of all, this season, to go down. "The moon had come in on dry earth," as the harvesters expressed it; and with implicit faith in the moon, there would therefore be no rain. For once in a way faith was not misplaced: there was great heat, which ripened wheat and oats and barley too quickly, left the straw short, and covered the turnips with fly.

It was too hot in the day to go far—that is, for those in life who can choose 150 their own time. So the dog and the man took their walks late, and prolonged them to the hour when the ruddy moon rose solemnly into the sky over the woods and set out on its low, summer curve to the west. Daylight lasted long after the sun went down: a hot glow spread gradually northward, and what with the light in this direction and the moon at full, only those two other worlds, Jupiter and Venus, were visible in the cloudless vault above. This was the time of day to be abroad, but, oddly enough, the hour when many were indoors. There was some excuse for the harvesters. They had been up with the sun: by half-past seven it was time to put the self-binder to bed in the field; by eight, or soon after, many were in bed themselves. Men and horses had sweated much, and had had a long day.

It was on an evening such as this that 151 Murphy had his first lesson in working to the hand, for Job's remark had given rise to a train of thought. Education was of course everything. Those who lived on the land should be educated in the things of the land; should learn, if not its deeper wonders and mysteries, at least its simple lessons and what lay at the back of these. It was in these fields and over these breezy downs that thews and sinews were to be braced, health and strength gathered, souls cleansed, if so be that the ways of the man were straight and true.

Here was God's work always visible, from the wonders of the growth of the seeds to the coming of the music of the rains that washed the air and made the land sing with life. Here was always visible the infinite power of small things, beauty unstained, Nature's laws always in full operation—the triumph of good work, the smothering of that which was 152 ill. Here in these very fields had been gathered the strength of arm that had stood the country in good stead, when the drums beat and true men were wanted beyond seas. That seemed to be more as it should be. And so it may be yet—that is, when the craze of a day has passed, and the men of the land come back.

Education would do it. Some hearts would be bitten with the old love, and learn to forget the new. But the education must be true and not false, in tune with the life that shall be; not cramped and with little connection between it and the field of labour that lies ahead. Uniformity is often but to bring down to one dead level, to crush true liberty and freedom, to force unnatural growth, and to give this a trend untrue. Education on such lines seems curiously false to many minds, as well as stultifying.

Scot, who had no appearance of a 153 sheep-dog—that is, as his class are generally portrayed in coloured prints—might possibly have been brought up as a water-spaniel, or he might have been the darling of a semi-detached villa and have learnt to walk drab, unlovely streets without endangering his life: it is all a matter of education, fortified by environment. As it was, he was brought up with a cottage for a home and learnt the mysteries of sheep, the tending and the care of them, what the stretching of limbs meant, no less than freedom and free air.

The life was a hard one, no doubt, in one sense. Sometimes there were short commons: there was much bad weather to be faced, when his master was clad in strange clothes and wore a sack like the hood of a monk over the top of his weather-worn cap, and he himself was glad to get to the shelter of the hut, where the stove was burning: there was the wet, 154 when all alike were mud-smothered: there were the biting winds of March. But there came the glad spring and the long summer days; the one gave a flavour to the other and created a love for both, and deep down in the heart where that love burnt bright was the pride of his calling, the honour of tending sheep. Soft jobs were not for men—or manly dogs.

Of course Murphy could not be a sheep-dog; that is, unless Job Nutt had a mind to make him. Then, of course, he would have had a proper schoolmaster, and been brought up to things among which he had been born and bred, while lookers-on beheld a novel kind of sheep-dog. As it was, however, his master owned no sheep. Yet, seeing that his lot had not been that of some—to walk the streets for exercise, or to lie in the cramped garden of a villa in a town—it was only right he should learn all that he could, and that his education should partake of the fields and the upland downs around his home.

As to whether it would have been possible to have trained him to the streets at all must now be left among the things unknown. The impression remains that, seeing he never grasped the desperate dangers of the modern road, his life, had he been so foolish as to forsake the country for the town, would probably have been limited to hours. For a better, freer life he was fortunately born, and he certainly never threw this chance away, but made the very most of it, and came to great happiness thereby.

Of course it took time; but a beginning was made in those halcyon, summer days, and the art of working by the hand gradually brought to some perfection. No little of this dog's gladness in life was centred eventually in this accomplishment, and he was never happier than when at practice. The education began by teaching him to lie down at the command—"Stop there," and then in leaving him behind for gradually lengthening periods. So well did he know these words, that he would act on them instantly, and in this way once lost his walk by a slight misunderstanding. An explanation of the method was being given one day, when walking with a friend. The opening words were of course used. Some time after the dog was missed, and it was not until steps had been retraced for a considerable distance that he was found, lying where he had first heard the words and looking a little shy.

The next proceeding was to start him, and then to stop him, till by degrees he came to understand the movement of the hands or arms. In this way it was possible to send him to great distances, or move him to right or left, much after the manner in which we who are soldiers move our men. When a hand was uplifted high, he would drop at once, so that nobody would think that there was a dog within a mile: he might be lying in rough grass where the ragwort was high, or the wheat, as they say, was proud, and be himself invisible. But he could see well enough with those bright eyes of his, and the moment the arm was waved he was off with a stride of two yards or more, circling round and making the valley ring to his glad bark. He always entered into the whole fun of the thing, and looked upon it as the finest game that had ever been invented.

"Ah, well," remarked Job as he watched, and Scot gave tongue for very jealousy—"ah, well, I allus liked that dog." 158

And so did every one.

With each little addition to the sum of knowledge he possessed, master and dog grew closer to one another. It is always a moot point whether our dogs consider they belong to the family with which they live, or whether they do not regard the matter the other way about, and judge that the family belongs to them. In Murphy's case there is no shadow of doubt that, so far as his master was concerned, that master most certainly belonged to him. At first, the position had been different. There was reason for that. But even the reason had now apparently passed out of mind: injustice had doubtless been forgiven, and what was far more wonderful—or rather, would have been, had man been in the case and not a dog— had also, so far as could be seen, been totally forgotten.

So completely had confidence been 159 won that anything was permitted, even to the playful brandishing of a stick. Sticks were things to play with. They had no relation to punishment at all. Besides, was not life a state to be enjoyed, and as happy as the day was long? And had he not taught his one great friend no end of facts of which he had hitherto been desperately ignorant?

It was all very well for Him to say that he had educated and trained this dog. The dog had all the while been training Him. It was all very well for Him to think in his heart that he had given this dog happiness in life. Happiness had in a measure also come back to Him. There had been, in more than one direction, a strange parallel between their cases, and as this had made itself felt, it had bound them both more closely together. They were now not only never apart, but they were of one mind in other 160 ways as well—in joy of life as they found it under the sky; in the happiness of comradeship as they learnt to rely on it—indoors and out; in the deeper meaning of friendship, with the trust and undeviating truth that friendship claims; in the faith that the one had always in the other, through the good days and the hard.

Those who watched were often overheard to say, "The dog has taken charge of the man." And so he had, to a certain degree. He had learnt his master's habits exactly. He knew the time of day by the striking of the clock; and, morning after morning, at a particular hour, if this master, with his funny ways, delayed his going, he would get up from his familiar corner and come and stand and fix him with his eyes. Or, if this failed, would come, gently, closer, and lay his chin upon a knee, and make him lay down his work and come out for the regulation interval. In the longer marches of old days, there were halts in every hour. Come out! Come out! New strength and new ideas are to be gathered outside; you will grow stale in here, whether you choose to practise this art or that. Houses are well enough to sleep in and to give shelter; but it is the heavens that give strength, and it is God's heaven that somehow, if only feebly, must get itself reflected in man's work.

So, in another instant, these two would be out together; the one going as far as tether would allow; the other doing what was yet another of his joys in life, and that caused such fun and merriment to lookers-on—the hunting of birds. Of that he never tired on the longest or the hottest day. Blackbirds gave the finest sport of all, as they generally flew only three feet above the ground. He knew their note at once; but probably the laugh of the green woodpecker vexed him more than most, while he certainly regarded the mocking notes of cuckoos as insults to himself. Of birds of various kinds he caught many, young and old, but was never known to hurt a single one.

The most remarkable of his exploits in this direction was when he found himself at one time by the sea. It was a lonely coast, where great crimson cliffs rose sheer out of the sand, their ledges, here and there, covered with tamarisk, gorse, and shaven thorn—right to their very summit three hundred feet above, from whence the moors stretched far away inland. A heavy surf beat there at times, setting these cliffs echoing in such a way as to make speech difficult. On these wild days it was well that this dog had learnt to work so perfectly by hand, for he had no fear of the rollers, and the wonder was that he escaped from being drowned.

At the bottom of the whole fun of this new situation lay the fact that these cliffs were inhabited by innumerable gulls. To catch one of these was Murphy's aim, and often was he washed out on to the sands in a smother of spindrift, in his mad eagerness to attain his end. The herring-gulls were the finest sport of all, with their constant melancholy cries—"pew-il," "pee-ole," or their hoarser note of warning, "kak-k-kak"; their bodies two feet in length; their spread of wing no less than four feet four. For months he chased them, till at last some must possibly have known him. It was perhaps on this account that one of them was not quick enough in getting under way on one occasion. Murphy flung himself into the air and got him; and not only got him, but brought him along, with the great wings beating the air about him, so that the dog was scarcely visible for the bird. It was the old story again, of the hare in his earlier days, for the gull was not harmed, and when liberated flew out to sea, with the cry "pew-il," "pee-ole" flung back from the waves as he went.

"I never thought to live tu zee the like o' that," remarked a longshoreman passing at the time: but then he was a stranger to Murphy, and also to his ways.

What happiness was to be had in life; what sport and splendid fun—sport all day long; fun without end! Did not the morning begin with a game?—the dog lying down in one corner of the hall, fixing his master with his eye as he appeared, and then, after pausing a while as if to say, "Are you ready?" launching himself full tilt, till he was brought up in a final leap against his master's chest, full five feet from the ground. Of course the whole hall was in a smother every time, with mats and rugs all out of place upon the slippery floor. And then the noise! The only thing was to leave the house and work off some of the steam out there.

No dog with a particle of nervousness or hesitation left would do such things as that. But he only did them with his master. When with others, report had it that he was a different dog, with no taste for hunting or for chasing birds—a dog, in fact, that invariably got into one room and lay there alone, unless he changed his place for the mat by the front door.

Of course He would come back. Folk always did. There could be no break in this friendship: it would last for ever. He had heard his master count the years: "Four"—that was his own age—he knew that much; and from four his master would count up to ten; then hesitate; then say "eleven"; then hesitate again, and remark, "twelve—perhaps: yes, little man; you'll see me out—easy!"

And those who watched and looked on added this to what they had said before, "What *will* happen, if anything happens to that dog?"

It was a funny way of putting it, but the remark was always met, in reply, with, "Don't let us meet trouble half-way, or make a circuit of the hills to look for it;

"'Fortis cadere, cedere non potest.'"

XI

The roads were deep in snow. The fall had begun two hours before light; gently, and with large flakes—the presage of what was to come. Snow was still falling in the afternoon; but now the wind had sprung up, and each large flake was torn into a dozen as the wind played with them, driving them upwards like dust, then catching them and sending them horizontally and at speed over the ground, till they could find a resting-place in some drift that was forming on the north sides of fences, or peace beneath the brambles of some ditch.

An hour or more before dark the wind increased, and was blowing a whole gale. What fun to be out in that: come on!

It was not long before man and dog 168 were away. The roads would be safe on such a day as this; so, for once, the two trudged along till they overtook two waggons. How big they looked in the smother, each with its team of three—a pair in the shafts, and one more ahead as leader. Talking was difficult, or well-nigh impossible; but at least they could join the men, and shout a word or two at times.

On the weather side the great horses looked twice their size, plastered as they were with snow, their manes and the hair about their huge feet all matted with ice. But on the lee they looked different animals, for their coats were darkened, being drenched with sweat: it was with difficulty that they kept their feet, and their breath came heavily through their nostrils as they struggled on.

Not that they had a heavy load to draw. The waggons were empty. They 169 had come in with a full load in the morning, intending to bring coal back. "But how was 'em to do that, in weather the like of this; or on roads same as these here? Nay, nay," shouted the rearmost carter, "we's for getting home, empty or somehow, if so be as these here can keep their feets. The road below the snow is ice, I tell ye—just ice; and, what's more, Fiddlehill lies just ahead for we." The last words were punctuated with the crack of a whip like a pistol-shot: all talk was dropped after that for a while; the wind was growing fiercer.

Both waggons were painted yellow, picked out with scarlet; but the paint that had looked brilliant in the sun of the harvest days looked tawdry and dirty now against the snow, and every patch or scar of rough usage was easily discernible. Now and then the wind came with a savage gust, carrying stray straws out of one 170 of the waggons, though snow was collecting on the floor: on the other, the cords of a tarpaulin, indifferently secured, were smacking the yellow sides like a lash. Some of these sounds did not suit Murphy very well; but he had found out the best and safest place, and was making his way as well as he could, sheltered beneath the rearmost waggon and between the tall hind wheels, whose rims and spokes and hubs were hung and bespattered, like all else, with snow.

It was true that he looked like some other person's dog, with a white face and whiskers. But his master was white, too, from head to foot; what recked it!

In another hour or less darkness would have shut down on the world, though such a term as darkness was only relative on a day when it could never have been said to have been light.

When the open was reached, the snow, 171 broken into hard flakes, whipped face and ears like nettles. Murphy was the best off of the party, save when something had drawn him from beneath the waggon, and he was having a game with the snow on his own account. Great wreaths hung to the fences, or stood out in ledges where the banks were high. The sky, or rather the whole air, was lead colour, and all distance was blotted out. Flocks of crazy, distracted birds flew close by in great numbers, for the most part finches and larks, with here and there a fieldfare or two, their breasts and underwings buff colour. Then came a flight wholly made up of buntings, whose brilliant yellows looked deep orange against the leaden grey that shrouded all.

There was no end to the great host. They were all going one way: they made no sound but the swish of wings, and uttered no single note: they passed at speed 172 as though in fear, yet all the while in obedience to the supremest law of all. To the southward there would be protection; life there would be preserved: here it was impossible—for birds. "Keep low; press on!" Victory shall be to the strongest: the weak shall fall in this pitiless wind, and the snow shall cover the dead, but in the end there shall be a better life for some. "Keep low; press on!"

There was something weird in such a sight as that: there was something weird also in the sound of the wind. It came sweeping over the fields, tearing with angry gusts at the snow-laden briars in the fences, and passing on with a moaning sound into the dark of the approaching night.

There was no sign of human beings anywhere. Familiar objects had all changed their character, though it was only by these that whereabouts could be 173 told. The remains of a hay-rick by the roadside suddenly showed up out of the mirk, with white top like some great ghost, its blackened sides flecked here and there with snow. In the hot days of June two here had seen it built; and, later on, watched the trussers at work on it, when the price of hay had gone up, and farmers could make a few pounds. But that job, like most others, had had to be abandoned now.

Why, here was the great stoggle oak by the pool, on whose limbs in former times, tradition had it, many a highwayman had swung! The storm to it was nothing: it had weathered so many: the world was a fair place; but life was full of tests as well as trials. "Heads up! Bear yourselves like men," its limbs seemed to roar in solemn, deep diapason. "Heads up!—there is a haven for all ahead!" 174

It was fifty yards further on before the voice of the oak was lost. But as man and dog worked further still, for very joy of the wind and the snow and love for the elements at their worst—the horses struggling, the waggoners calling to them loudly and urging them to put their best into it, with many a crack of the whip—there suddenly fell a lull, and for a moment there was peace. And just then, up from the valley, there came other sounds—the larch and the firs down there were sighing out a tune to themselves, being partly sheltered by the hill.

It was time to turn back. There was a lane in the direction of those last sounds: home could easily be reached that way, and, likely enough, with the set of the wind, the roadway itself would have been swept almost bare.

The waggons were lost to sight in a 175 moment, though the woody rattle of the axles could still be heard: snow was falling heavily again: the cold was becoming intense: the wind was now dropping altogether. A dead bird or two were passed, lying in the snow, claws in air and already stiff: a felt and a yellowhammer were side by side at the bottom of the hill. It was like the dead in gay uniforms, lying scattered after an action. A little further on there was a blackbird, to Murphy's very evident glee. He found it at once, and was for carrying it home; it was still warm. But this was no time for fooling. It was already dark and growing darker; the proper thing to do was to keep together and make for home. Travelling was none too easy, even for tall men, and really difficult for dogs in places.

At points where field gates opened on to the road, drifts had formed two feet 176 in depth, right across the way, and it was necessary to pick up the dog and carry him, though to the latter's thinking that was a silly thing to do. Time was, when his master had had to do that; but he had then been no better than a child in arms. Now he was a man, and had come to man's estate, and, furthermore, had learnt what life was, with its hours full of health, and crammed with fresh adventures and experiences, as, of course, it should be. His muscles were hard and flexible as steel, his heart strong with life, his brain quick to learn whatsoever his master thought best that he should know. Health, strength, what happiness it all was! The neighbourhood of those waggons had been rather depressing, and the crack of those whips somewhat disconcerting; but he did not stop to reason why. It was enough that he and his master were together. The past might look 177 after itself, and so might the future; this was the all-sufficient present.

A deep silence reigned in the valley; even the larch and the firs had given up their songs. There was the scrunch of the foot at each step, and now and then a rustle in the hedge, as a bramble became overweighted with snow and dislodged its load into the ditch, or last year's leaves, still clinging to some oak, rustled and were still again. Otherwise the world was dead or asleep; it made little difference which.

A cottage was passed further on, and a chink of light from a candle within showed that the snowflakes were still falling fast. This way would be impassable by morning. At the turn of the lane voices were heard. They were some way off; but it was easy to recognise that they were those of two men talking. Presently the voices became more audible. 178 It was too dark to see who the men were as they passed: at night, when snow is falling, those met are up and gone by almost before their approach is realised. There was just time for a "Good-night," with a "Good-night to you, Sir," in reply.

For an instant there was silence: then the men began talking again.

"Bless the Lord! — did you see who that was, Tom, and on such a night as this!" remarked one.

"Don't know as I know'd un."

"Not know un?"

"Why, bless the life on yer — that's Him an' his dog!"

"There, was it now? Him an' his dog, for sure. Carrying un, wus he? Like un."

"Ah — allus together, ain't 'em?"

"For his part, he don't seem to have much else."

It would be well to get on, and not to 179 stand there gaping into the darkness, listening to what you were never meant to hear. The truth of the old saying generally holds good; and sometimes words accidentally overheard in such ways are fixed in the mind for life. These last were like a stab.

"Don't seem to have much else?" What did the fellow mean? How invariably lookers-on misjudged! What a mistake it was to pass judgment at all — on anything or anybody!

"... Much else ... much else...?"

The road was less deeply covered here. The dog was heavy: a few yards more and he was put down. As the journey was resumed, he took to playing in the darkness, and, in his winning and affectionate way, with the fingers of his master's hand, as much as to say, "Thank you: we are together; the rest matters little." 180

"Him and his dog ... much else ... much else...?" The words kept time with the footfall.

How dark it was!

And cold—the thermometer marked minus 1°.

XII

A summer night, and the heat the heat of the dog-days. The tram-cars had stopped running long ago; the streets were quite deserted.

Not long since, the clock set high in the tower of St. Giles' had chimed three-quarters; and now it chimed the hour, and wearily struck "Two." Then other clocks also awoke to their duties, and, not possessing chimes, repeated the latter information in various keys, from far and near. It was all very sombre; and the smell of the streets very unlovely.

It was Bill's turn to be up that night; at least, they said it was his turn. As a matter of fact, he had been up three nights running, and at least ten in the last eighteen, for this was no ordinary case, and the credit of the firm was at stake. Not that he held the dignity of being a member, much less a partner, of the firm; but he had worked for it, he would often tell, and with no little pride in his voice—"worked for it thirty-two years, come Lammas; and that wus a very long while."

To Bill, and the few remaining, or still discoverable, like him, the firm's credit was his; and the firm should never find its confidence misplaced so long as Bill Withers could walk on his two feet, or aid some suffering creature. Those were his sentiments. Then, of course, this Bill had a soft place in his heart for animals generally, though the softest place of all was unreservedly retained for dogs.

"They wus human; well, a sight better than human, as any one might see humans at times";—that was the way he put it. "And there warn't a mossel o' doubt about it, no matter what nobody said."

At that, his mates in the yard thought well to let the matter drop.

"That there Bill has his queer hideas abaht most things; better leave him to hisself," they remarked, with a twist of the mouth, and passed on.

Bill had a habit of speaking his thoughts aloud, especially when up at night. He found company in the habit, and was employing his time in this way now.

"Two o'clock. Another half-hour and he'll have to have the soup, and then a little stim'lant. That wus the orders. Let's see. Tomorrow's Toosday. That'll make it three weeks since the master brought un back with him in his motor, all wrapped in blankets. 'Twas that ogg-sigen as saved him at the moment. But here—he's been fed every two hours, night and day since, any way. Well, well..."

There was a step on the cobbles of the yard. Bill looked round. "Mr. Charles"—as he called him—the head of the firm, was coming.

Five weeks before this Murphy had been taken ill. Nobody appeared to know what was the matter with him, except that he was restless, refused his food, and looked wrong in his coat. The very spirit there was in him misled others: he would hunt birds under the smallest provocation; rabbits were not animals to be given up so long as there was breath in the body; that finest of games, working to the hand, was to be played to the last day, for was it not the jolliest of fun for both, and did not his master laugh loudly when it was all over, and he skipped and barked and jumped himself, asking for just one more turn? It was only the chicken-hearted that gave up; life was to be lived to the very last minute, especially when so full of fun and happiness as his. If he flagged and was tired after these doings, it was only the hot weather: he would be all right tomorrow. So he was kept quiet for a week.

But the morrow came, and he was less full of life than on the day before. There was something evidently wrong; though advice was asked, and with little gain. His bright eyes had grown dull now, and he refused all food. It was time to call in the best opinion that could be had.

"Distemper. Pneumonia; and the heart also affected." That was the verdict. There was just a chance for him. It would be a risk to move him so far; but it was perhaps worth it, as treatment could then be followed properly: in establishments of the kind all animals were tended with as much care and skill as patients in a hospital.

So Murphy was taken away. How suddenly it had all come about. And now three weeks had gone by; and the dog still lived.

"How's he doing, Bill?"

"No difference to my mind, as I can see."

"We must save him, if we can, Bill. She was here again to-day, and said the dog was such a very valuable one that she didn't know what would happen if he died."

"I judged something of the kind," remarked Bill. "I've got a cousin, over their way: shepherd to Mr. Phipps—him as has Fair Mile Farm. You knows. He come in with him—'twus last Saturday's market—over some tegs; and he called in here, and I do believes 'twus to ask how this un here wus. Said he'd allus 187 liked un. Seemed to know all about un. Said as he and the gen'leman as owns un wus allus together; that he couldn't get about like some; and that he and this dog here was never apart, and seemed to hang together, curious ways like. They'd got some name for the two of 'em down in that part—so he says; but I a'most forgets what 'twus now."

"So I understand. One or two have been to call to ask after him, up at the office, and said much the same."

"Been here himself, hasn't he?" inquired Bill.

"Ay, yesterday. I told him he couldn't see him; or, rather, that if he did, with the dog's heart as rocky as it was, I would not answer for the result. He did not speak a word after that, except—'Do your best'; and went out."

"From what that cousin o' mine said," put in Bill, "I judge if he'd come in, it 188 would a-killed the dog right off." He was smoothing Murphy's ears as he spoke.

"I told him," continued Mr. Charles, "that two things were especially against this dog; one was his high breeding, and the other, his brain development. It's the last I'm most afraid of, though."

"Brain? Clever?" put in Bill—"I should just say he *was*."

" — And I told him that I had never seen a dog that was easier to treat; and that he was making a real plucky fight for it."

"That's true," said Bill, in a tone as if the words had been "Amen."

"—And that he was that sensible that he allowed us to do just as we liked with him; so good and patient that there was not a man in the yard that wasn't *glad* to do anything for him." 189

"True again," broke in Bill, with emphasis.—"Murphy," he said, calling the dog by name. "Whew! Another hot day, I judge; coming light afore long." Bill was looking at the sky.

"All against him; all against him," returned the other. "But there, I shall be downright sorry if we lose him now."

Bill shook his head. "See all as has been done ... and the telegrams ... and the letters, and ..."

The conversation of the two men was stopped by a low bark from the dog.

"Dreaming," said Bill; "does a lot o' sleep."

"Brain," said the other, listening—"I feared as much all along. It's all up, Bill."

Bill was down, and had got one of his hands under the dog's head.

The bark came again: only a very weak one; not enough to disturb anybody near. 190 It became continuous after that; grew a little louder; then gradually fainter.

Perhaps he was hunting birds, though it may be doubted. More likely he was working to the hand over the sunlit fields, in the glad air, with a full life all before him yet; and in the company of one whom he loved with his whole heart, and to whom, while learning constantly himself, he, a dog, had taught no end of things.

There can be little doubt that he was working by the hand. Of course he was. But the hand that was beckoning him now was from over the border—from the land where there is room for both the man and the dog, and where there shall be a blessed reunion with old friends.

The bark died away: Murphy was dead.

"Not five years; or only just," remarked Bill. 191

Both men heaved a sigh.

Day was breaking as they walked away together down the yard.

A few days later came this, written by one whose business it was to tend the sick and the suffering among animals; to whom their passing was no rare event; and who must have had many thousands through his hands:

"I am so very sorry; but it was really a happy release after the brain symptoms had developed.

"I can only say your dog won the affection of all of us here to an extent unequalled by any other patient. I think this was due to the very brave way that he bore his sufferings, his kind and amenable temperament, and his almost human intelligence. There is no doubt that this last increased the susceptibility of his brain to disease, and made recovery hopeless."

Two men were working their way slowly up the Dene. They were the shepherd, Job Nutt, and his second. And their dogs followed them closely to heel.

They had just set out a new bait for the sheep on the vetches lower down, and were making for home.

Violet shadows had stretched themselves out to their furthest over the red wheat, now rapidly ripening; soon they would fade out altogether, and the woods would grow blue. For the sun was touching the line of the distant hills, and the long day's work was done.

"Why, there goes Him," says one, pointing up at the down to the eastward.

"So it be," returns the other—"Him and his ... Oh ah! but I was a-most forgettin'. I allus liked that dog"; and Job Nutt waved his hand.

All knew it. Contrary to what is generally supposed, certain items of news circulate rapidly among farm-folk.

XIII

It was only a dog.

Perhaps so.

The fact does not forbid the familiar question that rises always at certain hours to the mind of man, and will continue to do so till time shall cease, whether his friend take human or only canine form in life:

"But his spirit—where does his spirit rest?
It was God that made him—God knows best."

In truth, there is no answer to this question—"Whither?" And thus it is that we are compelled to leave it according to our habit when we are at fault, and much as the poet leaves it here. In the case of the man, we think we understand. In that of the dog, our difficulty appears 195 to defy solution: it is no question of argument, assertions are idle, dogma has no place. On the one hand we have those principles that come to man's aid, but of which it would be unbecoming now to speak. The vast majority of Christian men are enabled to ride out the storms of life without confidence wholly giving way, and with the first of sheet-anchors fixed in what is felt to be the best of holding ground. When, however, we turn to the possible future status of the dog, there is no sheet-anchor, and the holding ground is indifferent. Yet, in considering the case of the man and the dog, we are not left without a certain measure of support equally applicable to both. The spirit definable as the immediate apprehension of the mind without reasoning—the spirit of intuition—aids us on either hand. "We are endued," as Bishop Butler tells us, "with capacities 196 of perception"; and these enable us to accept much that lies outside the actual region of proof, because our inner consciousness tells us that we are not altogether on a false track, and that truths, if half hidden, yet, of a certainty, exist in the direction in which we are making earnest search.

We necessarily suffer here, as always, from the tendency that makes the wish the father to the thought; or, in other words, we not infrequently shovel the unpalatable overboard, that we may lighten the ship, and ride out this or that squall without quite so much strain upon the sheet-anchor aforesaid. The majority of mankind believe, and will continue to believe, most staunchly in what they wish to believe. Yet this tendency on our part—visible as it often is in directions where we should least expect to find it—does not necessarily prove our beliefs false, while 197 it also leads us not infrequently direct to truths, however unorthodox our course may have appeared to lookers-on.

In considering, then, the question of the possible future existence of our canine friends, the dominant feeling is commonly this: We believe that a future, in great probability, exists for them, because we feel that not to believe this would be to turn the whole scheme of the universe, as we understand it, into one little short of nonsense. We do not stop to reason: such things are because they must be; they cannot cease to be without total disfigurement of the plan of our conception. Intuition points, and almost impulsively perhaps, in one direction. There is "an intelligent Author of Nature or Natural Governor of the world." Life is not made up of haphazards. Eventually there will be happiness in completest form: otherwise there would be injustice, 198 and of this, life, as we know it, affords little or no evidence. For happiness to be complete, there can be no question of the songs we are to hear being indifferently harmonised, there can be no rifts in the lute: in a state of perfection imperfections must necessarily be imperceptible.

With our narrow, human limitations we are driven to conclusions naturally circumscribed and coloured by those limitations. We are cognisant of the narrowness of the field of vision allowed us, and we are perpetually made aware that we are beating our wings against the bars; but we nevertheless accept this or that conclusion because it satisfies our souls, or we refuse to accept it because we cannot honestly confess that it does so. Yet, once again, behind both acceptance and rejection there is something further—that intuition and power of perception 199 that enable us to find satisfaction in inferences that we know lie outside questions of faith, but which we nevertheless feel to be true. And the very fact that we are enabled to derive this satisfaction and to feel that our conclusions have an element of truth in them tends to confirm us, rightly or wrongly, in our conjectures.

Thus we come deliberately to the opinion that dogs will have a place in the land over the border. Such an opinion may be a bold one; but there is reason for believing that it is somewhat widely held. We naturally tend to materialise when we build up our several pictures; but we sin here, if at all, in the best of company. The city that lay foursquare, and that is described to us in the vision in the Island of Patmos, was of pure gold, with walls of jasper and gates of precious stones, and had within it trees and birds 200 and many divers animals, and material things of greatest beauty, besides the figures of innumerable angels. The description could not have been otherwise drawn if it was to be grasped by the mind of man, even to a limited extent. So with ourselves. To conceive of a world with all the attributes of beauty yet without flowers is impossible. To realise a world full of music and song yet without birds may be possible, but transcends the powers of most minds. To attempt to believe in the happiness of a world where companionship is to be looked for and reunion is promised, yet where the companionship of dogs is denied, is to strain the belief of some to the uttermost and not improbably to fail.

"Nor," writes Bishop Butler in his immortal treatise, "can we find anything throughout the whole analogy of nature to afford us even the slightest presumption 201 that animals ever lose their living powers; much less, if it were possible, that they lose them by death: for we have no faculties wherewith to trace any beyond or through it, so as to see what becomes of them. This event removes them from our view. It destroys the sensible proof, which we had before death, of their being possessed of living powers, but does not appear to afford the least reason to believe that they are, then, or by that event, deprived of them. And our knowing that they were possessed of these powers, up to the very period to which we have faculties capable of tracing them, is itself a probability of their retaining them beyond it."

When Robert Southey looked for the last time on his old friend, Phillis — and there is a bitter difference on such an occasion between looking upon the young and the old — he tells how often in his 202 earlier days this dog and he had enjoyed childish sports together, and how, later on, when hard times overtook him, he found delight in recalling the faithful fondness of the friend in the distant home, and longed to feel again the warmth of his dumb welcome. Then, when the old dog is at last dead, and there has come a severance of these precious associations, he breaks out with:

> "Mine is no narrow creed;
> And He who gave thee being did not frame
> The mystery of life to be the sport
> Of merciless man. There is another world
> For all that live and move—a better one!
> Where the proud bipeds, who would fain confine
> Infinite goodness to the little bounds
> Of their own charity, may envy thee!"

When we turn to the first of all books, the dog certainly appears to receive harsh treatment. The term "dog" is invariably 203 one of reproach. Goliath cursing David asks, "Am I a dog?" Abner exclaims, "Am I a dog's head?" St. Paul refers to false prophets as dogs. In the Psalms the dog is found to be synonymous with the devil; in the Gospels it stands for unholy men. Evil-workers are dogs; a dog is the equivalent of a fool; nothing is lower than a dog, and nothing is to be more abhorred. Finally, there is that hardest sentence of all—"Without are dogs"; as though any hope for dogs was entirely forbidden. It is the same throughout: the depraved of mankind are dogs, and the very acme of possible reproach and contempt is apparently to be found in the use of this one term. Abandon hope;—without, are you who are dogs!

But is the use of this term "dog" to be taken literally? There seems to be ample evidence that it should not be. The very extravagance of the language raises a 204 doubt at once, just as the grotesqueness of the application of the term shows that the dog itself could never have been meant. St. Paul speaks of false prophets as dogs because of their impudence and love of gain—characteristics hardly to be attributed to the animal itself. The term "dead dog" was the most opprobrious to which a Jew could lay his tongue; when David endeavoured to convey to the mind of Saul that the persecution to which he was subjecting him was a dishonour to himself, he asked him whom he was pursuing; was he pursuing "after a dead dog"? If, as Horace has it, "death is the utmost boundary of wealth and power," it is surely no less so of pursuit.

Then again, in the Psalms, David writes, "Deliver my soul from the sword, my darling from the power of the dog"; in other words, the devil. All dogs are not good dogs, though all dogs are good dogs to their respective owners; but no dog can possibly be classed as we find him here, or as the very image and likeness of the most depraved and debased of mankind as we find him elsewhere. He is incapable of these sins; he does not fall into these errors.

What we have to remember is apparently this. The earliest mention of the dog in Scripture is in connection with the sojourn of the Israelites in Egypt. The dog was declared by the Jewish law to be unclean; and it is not improbable that the Jews were so taught to regard him in opposition to those taskmasters who, they were well aware, held him sacred. Thus the term dogs appears often as the reflection of a passionate and deep-seated hatred, apart altogether from the animal's uncleanness, and also from the animal itself. The word came in this way to be a useful one to hurl at the head of an enemy at all times, or by which to classify those who lived outside the pale of common, human decency. For such as these last there could be no hope, and the term as applied to them was judged to carry with it the bitterest stigma, just as it continues to do in the East to the present day. To be a Christian is to be a dog; to be a Jew is to be a dog; an infidel is a dog; and to be known as "a Jew's dog," or "a dead dog," is to have sunk to the lowest depths of depravity in the eyes of all men.

But the way in which dogs were regarded did not stop with Jewish edicts and Jewish opinion. When the ancient Egyptians made way for another type, and Moslems took their place, the dog, honoured before as has been shown, fell at once into an inferior position. The Moslem law took its colour largely from Jewish practice, and the dog was generally looked upon by the Mahomedan as unclean. He continues, as all the world knows, to be still so regarded. The dog, in the East, is at once tolerated and neglected: he may be slightly better than the pig, but, like that wholly unclean animal, he is a scavenger, living largely on offal and what he is able to pick up.

He is thus, for the most part, a poor creature, leading a poor life, and being often much to be pitied. That he should have any future prospects before him, seeing him as he is, might well be doubted. But this must also be remembered, that if he is in various stages of development in these far-off lands, and with little chance of betterment, he does not differ greatly in these respects from vast multitudes of men among whom he moves, whether they be white, yellow, brown, or black. The conditions of his life are little by which to condemn him, 208 just as they would be insufficient in the case of others. Moreover, all classes certainly do not so condemn him, or do they look upon him in quite the same light. By the Parsees, for instance, he is not regarded as wholly unclean. Many of them keep English-bred dogs, as also do some of the more Europeanised natives of other classes, treating them much as we do, though this is still uncommon. Hindus of good class and Mahomedans are found generally to avoid them; but here again many Hindus, and such a caste as Sweepers, will touch a dog without considering themselves defiled, just as a Mahomedan will often hold or take charge of a dog, though he be careful not to do so by the chain, or leather lead, but by slipping his *jharan*, or cloth, through the dog's collar, and handling him that way. In many Mahomedan villages the dog is found in numbers, the inhabitants 209 being glad of his services in shepherding their goats, though condemning him to live outside the house, even though there be likelihood of his being carried off by a prowling leopard.

In certain directions, therefore, the dog is seen to be at least tolerated. But there remains one other remarkable fact to be noted. No one can have travelled in the East, especially in Turkey, without remarking the way in which the dog is generally regarded. Yet, in spite of this, he is all the while certainly classed as supernatural, and by no less an authority than the Koran. His uncleanness must be recognised; but, on the other hand, how are his fidelity and courage to be overlooked? They cannot be. And so this unclean animal, from whom men shrink, lest by chance their garments touch him as they pass, is given, as already related, a position in Mahomed's 210 paradise, and, because of his character, is deemed worthy a special place in that land of supreme bliss. There is a chance, then, for the outcast here.

It is time to look at the dog himself a little closer, and see what characteristics he can bring forward in support of hopes that many human beings entertain on his behalf.

Here is a dumb animal that, long before the dawn of history, is known to have been man's close companion. Step by step, we see him advancing with those to whom he is linked, until he raises himself immeasurably above all other animals, and takes his place preeminently as the friend of man. No one of those from whom he originally sprang was known to bark, and no wild species does so. By and through man, the dog was 211 endowed with this means of expression, and was thus able to act as his more efficient guard. It is an established fact that the dog barks when in contact with man, and loses the power when separated from him. Such was the case with the dogs that were left many years ago on the uninhabited island of Juan Fernandez. The descendants of these dogs were found thirty years later to have lost the power of barking, and only subsequently regained it with difficulty.

The fact that the dog barks is not, however, the chief point. This peculiar gift has been developed into a language, for it is by those wonderful inflections of the voice in barking that the dog has learnt to make man understand his meaning. Thus, as we all know, he is able to convey, at will, a note of warning, to signal the approach of danger, to show his anger, his alarm, his joy, the spirit that animates 212 him in the chase, to make his appeal for help, to declare the need of succour. His bark has in these ways become his chief means of communication, quite apart from the howl, the whimper, the whine, or the growl; the "singing" that is associated with a pack of foxhounds baying at the moon; the "talk" that the subject of these pages possessed to such an extraordinary degree.

Then again, as he responded more readily to education, and acquired by degrees something of the civilising instincts that were affecting man, the dog became not only a trusty companion but a humble servant. Nor did he stop here, for, what was still more remarkable, he certainly came by degrees to reflect some of man's chief characteristics, as well as nearly all human passions. By association of ideas he developed memory. By his dreams and the various sounds he 213 emits in sleep, he is seen to possess imagination.

His wonderful power of scent is found capable of being turned to other uses than sport, and is even now not utilised in sundry quarters as it might be. Then, too, he habitually forms his own judgments, and these are usually exceedingly correct, as when he recognises an intruder, or arrives at what is right and what is wrong within the circle of his own domain. On many occasions he certainly gives evidence of a conscience and the possession of the rudiments of the moral sense. When he does wrong he frequently exhibits shame as well as contrition, seeking forgiveness, and being often distinctly unhappy till this is secured. So far does he occasionally carry this, that when he knows he has transgressed rules, he will come and make confession, his own honesty bringing upon him a punishment he would otherwise have escaped, or serving to declare what was not previously suspected by those about him.

But it is when we approach the higher qualities that the dog stands out in his true light. The best of his class naturally possess these in greatest perfection, but it is a fact that none are altogether without them. His instinct, his patience and subservience to the will of his master, his pluck and his courage, his fidelity that nothing seems capable of undermining, his trustfulness, his power of sympathy with man and with his own class, and, lastly, the touching and infinite depth of his love—all these are characteristics that occasionally put man to shame, but which make man always trust him more and more. In the face of his marvellous instinct, man is not infrequently struck dumb as he watches. A dog's patience is a thing to study, as well as one from which to learn many a fair lesson. His pluck and courage are almost proverbial. In many a case the odds against him seem not to make the slightest difference: he will fight on to the end; let his master only lead, he will follow to the death.

And it is here that his fidelity attains its very pinnacle. Faithful unto death! Again and again, in innumerable instances, he has shown his faithfulness long after the one he loved was dead. The dog in the mediæval legend that dug his master's grave, covered him with moss and leaves, and then watched there for seven years, until he died himself, has found many a parallel in real life. A well-known dog in the days of the Stewarts was still beside his master's tomb three years after the latter's death; and, in much later times, another dog, at Lisle, refused to come away from the 216 spot where his master lay, and remained on guard for nine long years, the villagers recognising his fidelity by building him a kennel and bringing him his daily food until he died.

And if an instance of the exhibition of grief on the part of a dog is called for, some will remember the little dog in the far-away Sudan. He was the property of the only officer that fell at Ginnis, and who had been in the habit of taking him everywhere. When his master was consigned to the sand, this dog was seen to be cowering beside the stretcher, looking even smaller than before; and, when all was over, he had to be lifted away from the edge of the pit, where he lay with his head hanging over the edge in an abject state of grief. He was only a dog, and a small one; but many a man, hardened by the experiences of a campaign, turned away his head at the sight. 217

Few can have been much in the company of dogs without becoming aware of their power of sympathy, the way in which they almost invariably show this to their own kind, and also especially to man. For a dog to be injured or ill is for others at least to leave him in peace; but with man they go much further, as they do in many directions where man is concerned. When Lazarus lay at the gate of Dives, alone and neglected, it was the dogs that came and licked his sores. So, too, in the hours of human adversity, somehow or other, dogs appear to understand, and act accordingly. How often the expression is heard—"They know!" The reason of their conduct and their actions on such occasions is entirely hidden from us, just as is that strange sense that dogs of highly developed brains undoubtedly possess—awe of the unknown, and that has made some conclude that 218 they have an inkling of the spirit world.

Many dogs are subject to fits of nervousness, though for the most part only in connection with things they do not understand or are unable to grasp at the moment. At such times the dog invariably seeks the closer company of his friend, man. On the other hand, the dog often understands the meaning of sounds when man is at fault and a feeling of uncertainty has been aroused. A glance at a dog, and the words—"the dog hasn't moved," are quite sufficient then to reassure the watcher, possibly out of doors on a dark night. Thus the one looks to the other for support and confidence, and a mutual spirit of reliance exists between both.

There is little need to say much here of the dog's power of love, for every one is aware of it, or may have been made richer by it in his life. The old saying of 219 centuries ago still holds good, and "the dog is the only animal in creation that luvs you more than he luvs himself." There are those who assert that all love is divine in origin. If this be so, and the dog could be considered to have a religion, then undoubtedly his religion is the love of man. We are brought face to face here with a passion that, in the dog, knows no limits, and that is apparently incapable of alienation. Faith, truth, love! What is to be said;—whence come these amazing powers; for what object could they have been created here? Perhaps the matter were better left where that other was just now. We can only seek the shelter that is common to us in such circumstances.

"He knows, who gave that love sublime;
And gave that strength of feeling, great
Above all human estimate."

Once again, for ourselves, there is no 220 definite answer. The whole question forms but one more problem added to an interminable sequence, and in the face of which the man and the dog are both dumb.

Yet when we look back, and ask ourselves, "Are all these for naught?" is it still man's province to be mute? Many further questions crowd up to the mind here, as they ever do in yet graver issues. In our weakness and our anxiety we cannot suffer our case to go by default, even though we confess our inability to answer the questions one by one as they appear. We can only turn away our heads and say, "Such things can *not* be." This close relationship cannot be cut off and cease for ever. This touching interdependence cannot be brought to a sudden and a final end. The sparrows cannot be cared for and the dogs cast out. In other words, living things 221 among animals, not directly associated with human beings in their lives, cannot, surely, be singly preserved and those which have won our love and loved us in return be lost to us for ever and condemned.

Is it possible that all these marvellous qualities and characteristics, gathered together into one dumb animal, are to pass away and to have no place in the larger circuit of life? Are all these consolations that this animal, and this animal alone among the so-called dumb, is capable of bringing—are all the influences for good that he is granted the power of exercising upon the mind, the spirit, and the very soul of man—to be accounted of no worth; to be merely so many items to be used up in the furtherance of a great scheme and plan; to be dissipated even as the mists of the dawn when the day shall at last break? Surely,—can such things be? Human judgment and human justice are for ever fallible, and rough expedients at best. But that other judgment for which we look, and that other justice upon which we are wont mentally to lean, cannot possibly be either one or the other.

Something, then, of our case may assuredly be left there. We cannot answer the questions; but, as we confront them, we yet cannot cut ourselves free from that spirit of intuition spoken of above, or cease to draw our several inferences. Continuity in Nature faces us at every turn. All things work together for the final perfection of the whole—for the final transcendent beauty and completeness of the whole. There is unity in all. Of that most are certain; and men walk therefore in good hope. There is mystery at every turn. There is no escape from it. There is ever the demand for the making of a good fight in the face of it. And there is promise of victory in the end on the part of One

"Who by low creatures leads to heights of love."

We are not all willing to accept such things. We do not all, in our march in life, require the same tools to win our way. Neither do we all look in the same direction—not for help, merely, but for those common daily aids that we gather, or that are gatherable, from the simple and the great, from the animate and the inanimate, from the stained as from the beautiful and the pure.

In writing of the death of an animal second only to the dog, Whyte-Melville asks this:

"There are men both good and wise who hold that, in a future state,

Dumb creatures we have cherished here below
Will give us joyous greeting as we pass the golden gate.
Is it folly if I hope it may be so?"

It may be folly. Yet the writer of these pages does not doubt it. And therefore, in the quiet corner of the beautiful home, when Murphy was laid to rest close by Dan, these words were cut upon his headstone, in faith and in good hope:

MURPHY

DEAR BOY

1906-1911

"Thou, Lord, shalt save both man and beast."

www.ingramcontent.com/pod-product-compliance
Lightning Source LLC
Chambersburg PA
CBHW031441210526
45464CB00005B/2289